Small Scale Wind Power
Dermot McGuigan

Published in Great Britain 1978 by

PRISM PRESS
Stable Court,
Chalmington,
Dorchester, Dorset.

ISBN 0 904727 26 2

Printed in Great Britain by A. Wheaton & Co. Ltd., Exeter

Dedicated to all at
The Centre for Alternative Technology *
an alternative energy exhibition
and research centre

* Machynlleth, Wales.

Contents

Introduction

I feel particularly good about what is going on in the world today. It is true that we are rapidly running out of oil upon which our whole economic structure depends, but that is no longer a worry for there is something much more important quietly developing behind the scenes. This 'something' is difficult to define but its momentum and enthusiasm is unmistakable and distinctive. One aspect of this movement is the urge to work closely with nature, to endeavour to tap the abundance of ambient energy, whether it be solar, water or wind power.

It is becoming increasingly obvious that whilst these new sources of power cannot provide as much concentrated energy as oil, together they can certainly fulfill all our needs. Various sources of natural energy are sometimes combined to provide for a domestic house. For example, wind power can light the home and power electric appliances, whilst solar energy can heat both water and space. However, it is possible to power the whole house direct from the wind.

There are basically two ways to approach wind power. The first is to use a small wind generator, with a capacity of 200 to 1000 watts, to charge batteries which in turn provide power for lighting and other electrical appliances. This is called a low energy system. Most such systems are efficient, 'energy conscious' and go together with the idea that electricity, especially the refined national grid type is too valuable an energy to put to use for space and water heating — at least not when these requirements can be served by other sources of power. Low energy wind systems can be quite inexpensive.

The second main use is to install a large wind generator to power all the usual domestic appliances or to supply space and water heating. The former system invariably needs a large bank of batteries and an inverter, both of which can cost as much as the wind generator itself! The exception to this is if a device called a Gemini Inverter can be connected between the wind generator and a nearby power line. Then, the Gemini will convert the varying voltage and frequency from the wind generator to the same as the power line. Power can be stored in and drawn from the power line when required. This avoids the need for batteries or an inverter. The electricity for space and water heating is never taken through the battery or inverter, but instead, is taken in a crude form direct from the wind generator. Such a system tends to be expensive to install, though once installed there are no further fuel bills for many years and in this remember that the average fuel price increase over the past six years has been 25% per annum! Moreover it is likely to continue at that rate, if not increase.

Whilst that is sufficient reason for investing in a wind plant now, it is not the only reason for the recent resurgence of interest in aerogenerators. It is a nice feeling to know that your source of power is in your own hands and not in the hands of some impersonal monopoly whose only concern is self-interest. Apart from being beautiful in themselves, windmills use a renewable and non-polluting source of power. They are fun to work with and it is a great pleasure to watch a windmill as it spins in response to a gust of wind. I have yet to meet anyone who, on first seeing a wind generator, has not come to a standstill and stared with awe, and then — after the initial surprise — flooded the owner with questions. I hope this book will answer many of your questions on wind power. Happy reading.

I am particularly grateful to the following without whose assistance this book would have been impossible.

Thomas Cabot
Robin Clarke
Peter Eaton
Bob Fletcher of Environment Canada
Jack Park
Ed Salter
Dr. Kuno Tichatschek

EVALUATION

1 Energy Needs

"The material *needs*
of human beings are limited and in fact quite modest,
even though our material *wants* know no bounds."

E.F.Schumacher in *Small is Beautiful*

Determining energy needs is a most interesting exercise. It involves a close examination of lifestyle and priorities. What is needed and what is merely a useless burden? Wind energy can power the modern all-electric suburban home, but at a very high cost. Assessing the suitability of ambient energy sources for home energy and then adapting the house to these new forms of energy is, to me at least, more a functional art form than a task. I am assuming here that many of those reading this book live, or intend to live, in isolated or rural areas. By ambient energy I mean the abundance of natural energy that surrounds the home. Even in the desert there is this abundance — solar energy. Solar panels can be used for water heating for at least half the year. A solar space heated home is a practicality, it is also possible to retrofit an existing house to suit the sun. Combinations of solar and wind energy are compatible. During prolonged winter cloud even the best insulated solar heated home will begin to feel the chill. But it is in winter when the wind blows most. A wind generator may be installed which will supply the demand for electricity throughout the year and contribute the excess of power generated in winter to heating. The University of Massachusetts has done a lot of practical work on this combination.

In areas where timber is plentiful, the blending of wood burning energy and wind power is a natural choice. Wood burning will give space and water heating. Cooking with wood is not as easy as gas or electricity, but to those who live an unhurried life it presents no problems. Not all stoves will heat water or cook. A back boiler is needed for water heating and hot plates for cooking. Some wood burners also have ovens. Again, a combination of wood and solar energy to provide water heating is appropriate, solar in summer and wood in winter. Water power may be available, and if so the costs of a small turbine and a wind generator should be compared.

But whatever the combination of ambient energy sources is, there will always be a place for wind generated electricity. Wind power can light the home, operate the radio, television and stereo, pump water, heat space and water, and perform a miriad of other useful functions. The cost of wind energy sys- is directly related to their power output, which in turn is to be matched by the power need. The table lists a selection of appliances and indicates the power each uses. The table is only an approximation, and each potential wind power user should complete a similar chart. This is easily done by noting the consumed power of all the appliances used and multiplying the power, in watts, by the time they are used.

Domestic 'gas guzzlers' would obviously need a large and expensive wind system to fulfil their requirements. Whilst those with more modest needs will find a small and reasonably inexpensive wind energy system suitable. Asceticism or any other form of restriction has no place in wind power. But working with the wind does tend to increase energy consciousness and lead to an understanding that doing more with less is ennobling. Surely, in summer, a solar clothes drier (a washing line) is better than a steaming, odorous electric drier, and is not a hand towel a quieter and more pleasant way of drying the hair than with a noisy electric heater? You get the idea, a close look at what you need or want and at what you don't. Is a vacuum cleaner really any better than a broom or brush? Are electric tooth brushes and shavers just a fad, fostered by the commercials, or do they make any real contribution to human life?

Table 1. Approximate power requirement for various appliances.

Household Appliance	Rated Watts	Monthly kWh
Air conditioner	1300	105
Blanket (electric)	170	12
Boiler	1375	8
Clothes dryer	4800	80
Coffee maker	850	8
Deep fat fryer	1380	6
Dishwasher	1200	28
Fan (attic)	375	26
Fan (window)	190	12
Food blender	290	1
Food mixer	125	1
Fruit juicer	100	0.5
Frying pan	1170	16
Grill (sandwich)	1050	2.5
Hair dryer	300	0.5
Heat lamp (infra red)	250	1
Heater (radiant)	1300	13
Hot plate	1250	8
Iron (hand)	1050	11
Iron (mangle)	1525	13
Oil burner or stoker	260	31
Radio	80	8
Radio transistorised	6	0.5
Range	11,720	102
Roaster	1345	17
Refrigerator	235	38
Refrigerator-freezer	330	30
Refrgierator-freezer (frostless)	425	90
Sewing machine	75	1
Shaver	15	0.2
Sun lamp	290	1
Television	255	29
Television (colour)	300	37
Toaster	1100	3
Vacuum cleaner	540	3
Waffle iron	1080	2
Washing machine (automatic)	375	5
Washing machine (non automatic)	280	4
Water heater (standard)	3000	340

Another example is the fridge – in winter many of them become heaters in that it is warmer inside the fridge than it is outside the house. If an 'outside fridge' is used there is not even any need to go outside to get the food, at least not if a well insulated hatch is provided opening into the fridge. I will never forget the first (and last) garbage disposal unit I used. Suffice to say that I have never since come across a more pointless and costly gadget. Freezer? Each to his own, but I have no use for one, I enjoy food in season and not out of season.

Now, please do not get the idea that I am preaching miserable poverty, quite the opposite for I believe in abundance, but I do feel that true abundance is to be found in quality and self-determination, and not in heavily promoted gadgets. By rationalising energy demand and reducing waste, mainly through insulation, a considerable saving can be effected in purchasing a compact and simple wind energy system. The energy bonanza, which President Carter has so ably pointed out, is due to end, or begin to end, about 1982 when there will be a world shortage of oil. We have the opportunity to choose energy independence now and avoid the brute force of future circumstances. One of the happy consequences of running out of fossil fuel is that it will bring us forward (and not back) to the land, to a closer and more understanding co-existence with nature, to living with the wind and sun.

Examples and Costs

Where a house is not connected to the national grid the choice of wind power is attractive, especially as overground power lines can now cost in excess of £3000 per mile. A good wind energy system will cost less. The cost of a wind generator is little more than that of a diesel generator. But with the wind there are no fuel costs, no fumes, and the sound of a windmill is far more pleasant than the noise of a diesel generator. Where there is a reasonable wind, a wind generator is far more economical per kWh than a diesel generator. In such cases the economic argument for wind power is based on a simple capital cost comparison basis. And if that does not come out in favour of wind power, then a look at the cost of grid electricity and diesel oil will invariably settle the matter.

Cost of fuel relative to 1971

Year	Coal	Fuel Oil	Natural Gas	Electricity
1971	1.000	1.000	1.000	1.000
1972	1.109	1.056	1.000	1.000
1973	1.242	1.578	1.000	1.033
1974	1.529	2.345	1.075	1.061
1975	2.184	2.907	1.466	1.370
1976	2.322	3.078	1.607	2.031
Average annual increase	26%	41.4%	12%	20.6%
Overall average annual increase		25%		

Table 2. Cost of fuel relative to 1971.

There is no escape from the above figures, except by using inflation free alternative sources of energy. Do not be lulled into a false security by the apparently low price of natural gas. When the world oil shortage, due in 1982/1985, arrives, the price of gas, the production of which is closely linked to that of oil, will rocket. The cost of electricity, generated from oil will also rocket. At the same time the demand, and therefore the price, of coal will also increase. That there will be a world oil shortage is now beyond doubt – it is the consequences of such a shortage that have yet to be fully recognised. A detached look at our Western lifestyle and how it is utterly dependent upon oil is a shattering experience. Even coal is transported by oil.

This is all leading up to my main point – is wind power economically competitive with grid prices, when fuel inflation is taken into consideration? The answer, particularly in good wind speed areas, is yes. This is certainly so for large wind systems used for heating only, or for use with a Gemini inverter. Low output systems with batteries and inverter do have a much longer payback period, but there again their capital cost is relatively low. Assumptions about the future rates of fuel inflation must be made in assessing the economic case for wind power. Take an example. Suppose your fuel bill for heating is now £400 per year. If there is no fuel inflation you will still be paying £400 in twenty years time, and during that period will have spent £8000 on fuel. But if, as has been happening over the past five years, the cost of fuel increases by 25% annually, your annual fuel bill after twenty years will be a staggering £27,700. What is more, your cumulative cost – the total spent on fuel over the twenty year period will be around £137,000.

Now let us take an example of, say, a 10 kW Elektro wind generator, used for heating purposes only. The cost in Switzerland today is about £4000 including the control gear. Add a liberal £2500 for importation, tower, erection, and that gives a total cost of £6500. The power could be fed into block storage radiators, or through immersion heaters to a wet radiator system, or into under-floor heating. However, the wind generator, costing £6500, will produce 15,000 to 30,000 kWh per annum in good wind speed areas. A well insulated house of between 1500 and 2000 square feet uses between 25,000 and 45,000 kWh per annum for space and water heating. The current cost of solid fuel, oil and off-peak electricity is just over 1p per kWh, and over 2p per kWh for daytime electric heat. Calculated below is the time in years it would take to reach break-even point on a £6500 investment in an Elektro, or indeed in any other wind generator, given various rates of fuel inflation.

Even if the Elektro cost were doubled it would still, in many cases, be an attractive investment. Try it out with a calculator and see. Only in the worst case, 24 years to break-even, is there any doubt. In all other cases investment in wind power is justifiable. Those who say that fuel inflation will never again be as bad as it has been over the past few years are entitled to their point of view, but it is the consensus of opinion among many who have made a special study of energy that

Time in years to break-even
(assuming a £6500 investment in a wind to heat system)

| | 15,000 kWh | | 30,000 kWh | Energy Saved |
	1p kWh	2p kWh	1p kWh	2p kWh Cost of fuel
Annual fuel inflation				
25%	11	9	9	6
20%	13	10	10	7
15%	15	11	11	7
10%	18	12	12	8
5%	24	15	15	9

Table 3. Time in years to break-even.

after 1979/80 fuel prices will again be subject to rapid inflation. True, we may have a period of calm until then, but in the long run that is of little importance.

The life of a wind generator is always a debatable point. However it can be said with reasonable confidence that a wind machine, well built, sited above any turbulence, regularly serviced, thoroughly overhauled once every five years, and in general well cared for, will have a life of at least 20 years and possibly 40 years or more. There are many fine examples of Jacobs direct drive generators still operating after forty years. Direct drive Elektros have operated continuously up in the Swiss Alps for over thirty years. On the other hand a poorly made machine, uncared for, may only last a few short years.

Low Voltage DC Systems

This type of installation is the favourite of the self-sufficiency and conservation minded, where wood or solar energy is used for heating. Direct heat wind systems, as above, have a high capital cost but are low on maintenance (assuming the owner does his own maintenance) whereas wood burning requires little capital outlay and high maintenance effort. Small DC wind systems are relatively inexpensive and require little maintenance. Such a system could use the Winco 200 watt charging one or two heavy duty batteries to power low voltage appliances. the cost would be approximately as follows:

Winco Wincharger 12V 200 W with 10 ft tower	£350
2 heavy duty 125 amp-hr batteries	£100
Voltage regulator	£ 40
Total	£490

That is the cost for the basic system. Rarely will the 10 foot tower be sufficient. One inexpensive way to increase the tower height is to add sections of 2 in. diameter guyed steel pipe, preferably galvanised. Added to that is the cost of the 12 V cable, and if the Winco is any distance away from the house, I suggest buying a 24 V or 36 V generator which will require less expensive cable. The usable output from such a system, in an average wind speed of 12 mph, is about 26 kWh per month, or 860 watts per day. Not a lot, but sufficient to power 12 V fluorescent or car bulbs, a radio and possibly a small black and white television. The cost per kWh of such a system is high, very high, but still it has many advantages over a diesel generator. I know of a well cared for Winco 200 watt now entering its twenty-fifth year of service, and another unattended one which fell apart in less than five years.

There is a whole variety of wind generators with outputs higher than the Winco and eminently suited to power low energy systems. Such wind generators are the Aero Power 1 kW, Sencenbaugh 500 W and 1 kW, Dunlite 2 kW, Elektro 600 W and 1.2 kW, plus many reconditioned machines such as the Jacobs 1.8 kW — all costing between £1200 and £2400 for the wind generator alone, further details and prices are given under the manufacturers' section. All these generators give a considerably higher output than the Winco, and will therefore give a proportionately greater degree of home comfort. Though costs vary considerably the following gives an indication:

Sencenbaugh 1000-14 wind generator
complete with control panel   £1820
70 ft. guyed tower with top section   £ 275
900 amp-hr 12 V battery bank   £ 800
         Total   £2895

In an average wind speed of 12 mph the usable power will be between 130 and 170 kWh per month. Again, when compared to the current grid price, this works out expensive per kWh. If amortised over twenty years the cost will be about 8p per kWh. But then again I believe that by 1997 the grid cost per kWh will be way in excess of that, and of course 8p is still very competitive with diesel costs, even today.

The only people likely to buy such a system are those who live in high wind speed areas. In such cases the energy output will be far greater than that shown above, thus causing a considerable reduction in the cost per kWh. Other potential customers are those who are presented with a one, two, or three thousand pound estimate for connection to the grid. In such cases it is almost foolish to pay such a sum and then on top of it be faced with quarterly electricity bills. Why not invest the capital in a wind generator and enjoy free power?

The next system works out much cheaper.

Gemini Inverter Systems

The Gemini, or synchronous inverter can be scaled up to more or less any size. But here we will take an example using a fully rebuilt Jacobs wind generator:

Jacobs 2 kW   £2000
Self-supporting tower second-hand, 60 ft.   £ 330
Gemini inverter (UK made)   £ 400
Installation and incidental costs   £ 400
         Total   £3130

This system will supply about 350 kWh in an average wind speed of 12 mph (about 250 kWh at 10 mph). Amortised over 20 years, and assuming that there are no charging differentials, this works out at the cheaper cost of 5p per kWh.

Large Domestic AC Systems.

These systems allow users to take advantage of all existing domestic equipment such as freezers, large refrigerators, coloured televisions, pumps etc. The main element which differentiates it from the low energy DC system is the inverter. This is an expensive piece of equipment, but it allows the use of all standard appliances. The cost of such an installation can be broken down as follows:

Elektro 6 kW wind generator and top tower section

	£3000
50 ft Rohn guyed tower	£ 600
Battery bank, 450 amp-hour 115 V DC	£1700
Inverter 115 V DC to 115 V AC rated at	
6 kW sine wave output	£3200
Stand-by generator 2 kW	£ 230
Incidental expenses	£ 500
Total	£9230

As shown above the 6 kW inverter is more expensive than the 6 kW wind generator! A 3 kW inverter of similar quality would cost about £1900, but a 1.5 kW inverter is only marginally less expensive. Delatron, (see Manufacturers) have recently announced a range of special inverters for wind power users, a 6 kW unit will cost about £2400. It is as a direct result of the above costs that I urge anyone considering the purchase of a wind generator to re-assess their energy needs. Waste is expensive, and so much energy is needlessly wasted. The estimated monthly output using this system in an area with an average wind speed of 12 mph is between 400 and 620 kWh. Taking an average of 500 kWh per month, the cost per kWh amortised over 20 years is about 7.7p.

Let us look, on the other hand, at the cost per kWh on a direct heating system. The £6500 10 kW system — as outlined in the 'Time to break even' chart — if amortised over a period of thirty years and with an assumed output of 1600 kWh per month, would result in an actual cost per kWh of 1.1p. This is a reasonable price, particularly when that price will remain stable for the next thirty years.

True, there are no maintenance costs shown in the above figures. The reason for that is the maintenance involved with most wind generators is minimal. A good greasing once every few months is all that is required. Bearing and generator brushes may need occasional replacement but these are quite inexpensive. Free-standing towers need no attention unless they are of very poor quality or subject to rust. Guyed towers do need attention. The guy wires have to be carefully checked regularly, and adjusted if necessary. Finally, those who have wind systems with batteries must see to it that the batteries are topped up with water when needed.

And so, once a wind system is set up and working well (and don't let go of the installer until it is working well), the maintenance involved is next to nothing — when it's done by the owner. Anyone who wants a manufacturer or agent to service a windmill, to come out and water the batteries or tighten a nut, should really forget about wind power or else be prepared to pay a price penalty. Those who have the where-withall to climb the tower and grease the generator, to generally care for and respect the machine, will find that it takes little of their time and will give good service.

Further on in this book under 'Wind Energy Systems' I discuss nine working wind generator installations which show the various uses which wind power can serve.

2 Site Selection

The one essential ingredient for the operation of a wind driven generator is wind, and the more of it the better. Those who live in an exposed area, on a hillside, by the sea or on an open plain, are generally aware of the force of the wind. Many such people feel, and rightly so, that the wind could be put to good use. On the other hand there are those whose homes lie shrouded by tall trees at the bottom of quiet valleys where the wind rarely blows, and when it does blow it tends to be a turbulent wind which causes havoc with a wind generator. Unfortunately those who have the peace of the valley are unlikely to benefit from the wind.

The selection of a favourable site has a bearing on two important aspects. Firstly there are the stresses on the wind generator as a result of air turbulence and secondly there is the energy output of the wind generator.

The type of air turbulence most likely to cause disturbance results from local obstructions such as trees, buildings, rock outcrops, etc. When a smooth airstream encounters an obstruction, the reaction tends to be a severe buffeting which causes sharp variations in the stresses on different parts of the propeller and tower.

In order to avoid these destructive effects, it is essential to erect the wind generator at least 20 feet above any obstruction within a 300 foot radius, preferably 30 and 500 feet respectively. The drawings give a good indication of where and where not to site a windmill. Good and bad sites should be fairly obvious, judging from the lie of the land. In case of difficulty, tie streamers to the tallest portable pole you can get. Place the pole at different sites and watch for the rippling and curling effects of air turbulence. Turbulent air tends to roll in circles as a ball. The best site for the wind generator is where the streamer holds steady.

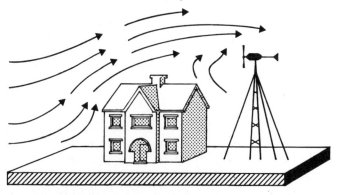

Figure 1a. The house is diverting the wind stream above the generator, which should either be raised or moved.

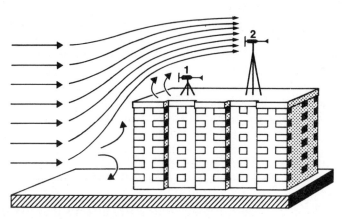

Figure 1b. Again, the building is diverting the wind stream from generator 1, but actually increases the wind speed for generator 2.

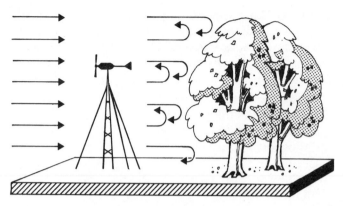

Figure 1c. Obstructions cause wind turbulence whether they are in front of, or behind, the generator, as here. The mast should be extended 20-30 feet above the trees, or sited 300-500 feet away from them.

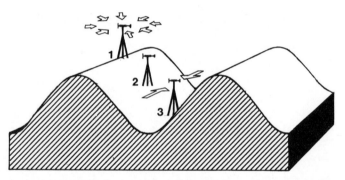

Figure 1d. Boundary considerations may rule out the best sites in hilly country. 1 is the ideal, and benefits from all wind directions. 2 is not recommended. 3 is a good site for wind coming from two directions only.

It is also best to site the windmill where it is well exposed to the main prevailing winds. This is important in those areas where the wind blows from only two directions for most of the time. If you are not familiar with the area and its prevailing winds, the Meteorological Office can supply wind direction data.

It is generally reckoned that where the average annual

wind speed is under 8 to 10 miles per hour (mph) that wind power is not a very attractive proposition. The power in such winds is low and it would need an enormous and expensive rotor to extract any appreciable quantity of energy. One of the laws of wind power is that power in the wind is proportional to the cube of the wind speed. This means that a 12 mph wind has *eight* times the power of a 6 mph wind and a 24 mph wind has *sixty-four* times the power of a 6 mph wind! This is an important characteristic, for once the average wind speed goes beyond the 10 mph mark to, say, 12 mph or more, then wind power really begins to look like an exciting possibility. To take another example, there is seventy-three per cent more energy in a 12 mph wind than a 10 mph wind. It is this additional percentage that makes all the difference in a wind energy system.

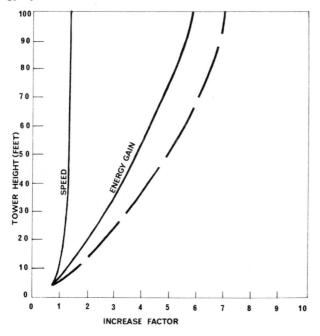

Figure 2. Increase in wind speed and energy with increasing tower height. Unbroken line is for flat land or sea and broken line is for rough terrain.

15

Wind speed increases greatly with altitude. So the higher the tower, the better it is for energy gain and turbulence free operation. Figure 2 shows both the wind speed and the energy to be gained by increasing tower height. The figure shows, for example, that there is about 3.5 times the power available to a wind generator at 40 feet, than at 5 feet, even though the wind speed is only about 1.5 times as great. It is true to say that the least expensive way to get more power from the wind is to increase the height of the tower. This tends to hold good from the point of view of cost up to about 100 feet, though there are many who prefer lower heights so that the machine is nearer at hand for servicing and 'keeping an eye on'.

Whatever the tower height, there is no doubt that the life of a wind generator left in turbulent winds will be considerably less than one operating in a smooth air stream.

3 Wind Measurement

The first step in planning for a wind system is to find out the average wind speed over the site in question. The most reliable way to do this is to buy or rent a wind measuring device called an anemometer or odometer. Where a low cost wind system is desired the expense of an anemometer may be avoided by the use of local weather records, or the use of the Beaufort scale.

The details of wind flow over any given area can be obtained from the Meteorological Office but the most that can be gained from their records is the general wind pattern for any given locality. One is then left to deduce if the same pattern is to be found on the site in question. This is done by comparing conditions at the recording site with those at the intended wind generator site. Most weather records are taken on open sites at a height of at least 30 feet above ground and 20 feet above the top of any obstacle within a 300 foot radius. If there is a distinct similarity between the two sites then one can be confident in using the weather records. There are other sources of wind data and you might try contacting local airports, schools and colleges, air pollution bodies and farming organisations.

More often than not there will be a difference in the lie of the land between the two sites. It is difficult, without a recording instrument, to judge the difference in terms of flow. An accurate indication of wind speed at any given moment

Table 4. The Beaufort scale of wind speeds.

Beaufort Number	Description	miles/hr	Effect on land	Effect at sea
0	Calm	Less than 1	Still: smoke rises vertically	Surface mirror-like
1	Light air	1–3	Smoke drifts	Ripples form
2	Light breeze	4–7	Wind felt on face, leaves rustle	Small, short wavelets, not breaking
3	Gentle breeze	8–12	Leaves and small twigs move constantly, streamer extended	Large wavelets beginning to break, scattered white horses
4	Moderate breeze	13–18	Raises dust and papers, moves twigs and thin branches	Small but longer waves, fairly frequent white horses
5	Fresh breeze	19–24	Small trees in leaf begin to sway	Moderate waves, distinctly elongated, many white horses, isolated spray
6	Strong wind	25–31	Large branches move, overhead wires whistle	Large waves begin with extensive white foam crests breaking
7	Moderate gale	32–38	Whole trees move, offers some resistance to walkers	Sea heaps up, white foam blown downwind
8	Fresh gale	39–46	Breaks twigs off trees, impedes progress	Moderately high waves with crests of considerable length, spray blown from crests
9	Strong gale	47–54	Blows off roof tiles and chimney pots	High waves, rolling sea, dense streaks of foam, spray
10	Whole gale	55–63	Trees uprooted, much structural damage	Heavy rolling sea, white with great foam patches, very high waves
11	Storm	64–72	Widespread damage (rare inland)	Extraordinarily high waves, spray impedes visibility
12	Hurricane	73–82		Air full of foam and spray, sea entirely white

can be gleaned from the Beaufort scale, see table 4. Two small and inexpensive wind measuring (but not recording) devices are manufactured by Dwyer Instruments (see Access).

Photograph 1. Dwyer hand-held wind meter.

The first is a small hand-held wind meter. The second, a wind speed indicator, can be mounted up to fifty feet away from the indicator panel. By using the Beaufort scale, or one of the two devices, a general assessment of wind speed may be made and compared with the daily records of a local weather station. Obviously if winds on a particular site are persistently strong and not too turbulent, that is reasonable justification for the purchase of a small inexpensive generator, like the Winco Wincharger, suitable for a low energy system.

Anemometers are not really all that expensive, a good one will cost around £60 (see Access). An anemometer

Photograph 2. Cup counter anemometer.

records on a counter the total run of the wind past its position. By reading the counter at the beginning and end of any period, be it a day, week or month, the average wind speed during that period can be accurately calculated. It is essential to use an anemometer if details of the exact wind power available is required. Also it is advisable to use one if the site is obstructed or not open to winds from all directions. The use of an anemometer in such conditions will help avoid any future disappointment with a wind energy system.

The way to assess wind patterns from anemometer readings is to compare the results with the detailed readings taken at a local weather station. The anemometer should be

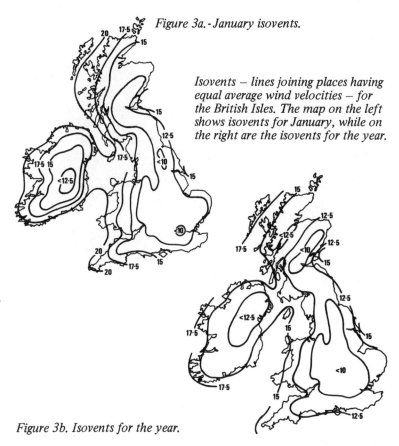

Figure 3a. - January isovents.

Isovents – lines joining places having equal average wind velocities – for the British Isles. The map on the left shows isovents for January, while on the right are the isovents for the year.

Figure 3b. Isovents for the year.

read daily or weekly for about three months. If readings can only be taken once a month it is better to record for six months. Next the readings for each period are compared with those of the weather station, and as a result a plus or minus relationship between the two sets of figures is ascertained. Once this correction factor is taken into consideration then the full wind spectrum on your particular site is known. The most important facts to be gained from the weather station records is the duration of given wind speeds, which in turn will give a good idea of what energy one can expect from any make of wind generator on that site. I say 'good idea' since even with the use of an anemometer, estimation of wind

21

energy is not an exact science. That is why, when choosing a wind generator, it is always better to obtain one with a greater capacity than you think you need.

There is a second way of estimating energy, suitable for use in locations where comparison with a weather station is not practicable, details of this are given in the next section.

Knowledge of seasonal variations in wind speeds is very useful in planning a wind energy system. Knowing these means knowing the energy pattern of a wind generator and planning accordingly. It is good that the wind blows hardest when we need it most, in the dark and cold of winter.

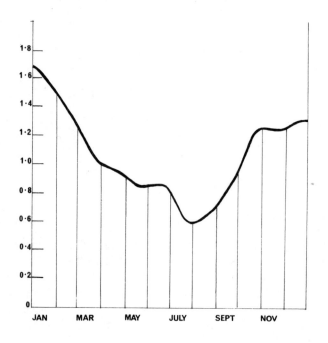

Figure 4. Variation in wind speed, shown as a fraction of the annual average.

4 Wind Energy Estimation

As explained earlier, wind power goes up as the cube of the wind speed, that is, double the wind speed and an eight fold increase in power is gained. There is a second rule, double the diameter of the propeller and the power obtained is increased by a factor of four. These two rules govern wind power, and are only limited by a third, as follows.

Energy is obtained from the wind by slowing down the air. The windmill cannot extract all the energy, otherwise the wind behind the propeller would come to a standstill. The Betz Law states that a windmill derives maximum power when the wind is slowed to one third of its initial velocity, in which case the power extracted is 0.593 of the maximum potential.

Based on the above, useful power in the wind may be expressed as follows:

$$P = 0.0031 \ A \ V^3$$

where P is in watts, A is the area swept by the propeller in square feet, and V is velocity (speed) in mph. This equation is based on an air density of 0.08 lb/ft^3, which holds good except for slight variations in high altitudes.

To avoid the trouble of assessing the swept area in square

feet, the same equation can be expressed more simply for most rotors as:

$$P = 0.0024 \, D^2 \, V^3$$

where D is the propeller diameter in feet. The two rules of diameter squared and velocity cubed can be seen in the above equation.

Using that equation, the maximum power available to a 16 feet diameter propeller in a 12 mph wind is expressed as follows:

$$P = 0.0024 \times 16^2 \times 12^3 \text{ watts}$$
$$= 0.0024 \times 256 \times 1728$$
$$= 1062 \text{ watts (or } 1.062 \text{ kW)}$$

But, inefficiencies within the windmill must be taken into consideration and deducted from the theoretical maximum power, shown above. A well made propeller (or rotor) will extract 70% of available energy. If there is a gear between the propeller and generator, its efficiency should be at least 95%. The generator itself should be no less than 75% efficient. It is safe to assume an overall wind generator efficiency of 50% (0.7 x 0.95 x 0.75 = 0.5). Accordingly, the actual electrical energy available may be expressed as:

$$P = 0.0024 \, D^2 \, V^3 \times 0.5 \qquad \text{or} \qquad = 0.0012 \, D^2 \, V^3$$

And so the actual output from a 16 foot propeller in a 12 mph wind is reduced from a theoretical 1062 watts as above to 531 watts.

Multiplying the average wind speed by the wind generator output at that speed will not give a true reflection of the average monthly or annual output. To arrive at such a figure correctly, the average wind speed must be broken down to the more detailed figures available from the Meteorological Office. For example, let us take a site some thirty miles away from a weather recording station. An anemometer is used for six months and the average wind speed is found to

24

be 12 mph at a height of 30 feet. From comparison, the wind speed at your site is 5% higher than at the weather station, which records an average wind speed of 11.5 mph. This difference is not very important, especially as it is in our favour.

However, a 5% correction factor is included in the calculations shown below.

No. hours per month	Weather Station Wind Speed mph	5% site adjustment	2kW wind generator output in kilowatts	Monthly output kWh
272	0–7	–	–	–
175	8–12	8–12	.4	70
188	13–18	13–19	1.1	206.8
64	19–24	20–25	1.9	121.6
24	25–31	26–32.5	2.0	28
6	32 upwards	33 upwards	2.0	12
				438.4 kWh

Table 5. Energy estimation for a 2 kW wind generator.

The number of hours that the wind blows at given wind speeds is supplied by the weather station. In the third column is the 5% correction factor for the difference between the two sites. In the fourth column is shown the output of a 2 kW wind generator at those wind speeds. The manufacturers of all wind generators supply such figures, and if they do not then they certainly should. The monthly output in kilowatt hours is easily ascertained by multiplying the number of hours by the kilowatt outputs in column 4.

The anemometer on this site was erected at a height of 30 feet, and it is at this height that the output of 438 kWh per month can be expected from the 2 kW wind generator. If the same generator is placed on a 60 foot tower, then we will have to revert to figure 2 to find the power increase factor for the new height. At 30 feet there is about 3 times the power than at 5 feet, at 60 feet, there is 4.4 times the power. By going from 30 to 60 feet on this unobstructed site, a power gain of 1.47 is achieved (4.4 divided by 3). Hence, the expected output of 438 kWh is increased to 644 kWh simply by adding 30 foot of tower.

Whilst the estimated energy from the wind is not an exact science, the above method is reasonably accurate. There is a quick and simple method which gives an approximate indication of the power potential on sites for which only the average wind speed is known. It is based on tests of 23 small wind generators. The curves drawn on figure 5 are not definitive but reflect a general pattern. The cut-in speed on the 20

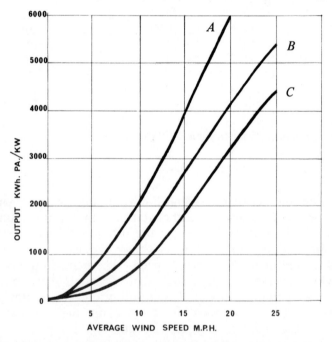

Figure 5. Annual output in kWh per generator kilowatt for various average wind speeds. Outputs for 3 wind generators with different rated wind speeds are shown. A has a rated wind speed of 20 mph, B—25 mph, and C— 30 mph.

26

mph line is 10 mph, on the 25 mph line 13 mph and on the 30 mph line it is 17 mph. The shut down speed for all three lines is taken to be 60 mph.

Let us return to the 12 mph average wind speed example used with the first energy estimation. The 2 kW wind generator used in the example had a cut-in speed of 8 mph and a rated windspeed of 25 mph. Using figure 5 an annual output of 1700 kWh per kilowatt can be expected for an average wind speed of 12 mph. Output from the 2 kW generator would, therefore, be 3400 kWh per annum or 283 kWh per month, to which must be added 70 kWh as the cut-in figure on figure 5 is 13 mph, whereas the 2 kW cut-in speed is 8 mph. So, using the graph, we arrive at a total of 353 kWh which does not compare too badly with the original of 438 kWh.

The whole question of assessing wind generator output from an average wind speed is now being made easier by manufacturers who are beginning to include such estimates in their catalogues. There is a reluctance on the part of some manufacturers to do this. They believe, and rightly so, that the characteristics of individual sites have an important affect on the wind generator output. But this need not stop them reporting the actual output of a windmill and describing the site characteristics so that others may make allowances for differences.

It should be mentioned here that output from the generator is not the final power available for use. Unfortunately, there are transmission line, battery and inverter losses to be deducted. These, and other wind energy systems, we will discuss in the next section, but first let's take a look at the various types of windmills which can be used to turn a generator.

5 Windmills

Strictly speaking, a 'windmill' is used for milling (grain grinding, etc.) and not for the generation of electricity or water pumping. But, being short and evocative the word tends to be used to describe what is properly called any one of the following: wind generator, wind driven turbine, windplant and wind machine, to mention but a few. There are basically two types of windmill:

Horizontal axis where the propeller (or rotor) on a horizontal shaft or axis moves in a plane perpendicular to the direction of the wind. This type includes the multi-blade, four arm and high speed propeller types.

Vertical axis where the rotor on a vertical shaft or axis has its effective wind catching surface moving in the direction of the wind. This type includes the more recently developed Darrieus and Savonius rotors.

Horizontal Axis Machines

Propeller driven, horizontal axis windmills come in all shapes and sizes, but there is one characteristic which differentiates the various types, that is their tip speed ratio. The tip speed ratio is the ratio between the propeller tip speed and the wind speed. This ratio can range from 1:1 for low speed mills, up to 8:1 for modern high speed propellers. Some recent reports indicate a tip speed ratio of up to 20:1 for single bladed propeller designs. Mills with low ratios are mainly suited for low speed purposes such as water pumping and other mechanical uses. High tip speed propellers are suitable for generating electricity.

$$\text{Tip speed ratio} = \frac{\text{Speed of rotation of blade tip}}{\text{wind speed}}$$

$$= \frac{2 \pi R \text{ rpm}}{88V}$$

Where R is radius, rpm is the revolutions per minute of the propeller and V is the wind speed (velocity).

Example: a six foot diameter propeller rotates at 700 rpm in a 23 mph wind, the tip speed ratio is

$$\frac{2 \times 3.14 \times 3 \times 700}{88 \times 23} = 0.65.$$

A tip speed ratio of 6.5:1 means that in a 23 mph wind, the propeller tip travels at 150 mph. If the ratio was 1:1, the propeller tip speed would be 23 mph.

Low speed high torque mills generally have propellers with high 'solidity' ratios, that is the area they sweep is 'solid' with blades or sails. They usually have a low power coefficient as the multitude of blades create air turbulence and cause a negative drag effect. Nevertheless, they are safe, suit the mechanical purpose for which they were designed and as a result of their high solidity ratio, they produce power at low wind speeds. Modern aerofoil propellers have a very low solidity ratio. They produce little or no power in windspeeds under 8 mph, but they make excellect use of high wind speeds where the really useful power is. Their high tip ratio makes them ideal for driving generators. Indeed, the problem faced by the manufacturers of very large (100 foot plus) propellers is to design them with a capability of holding together at the speed of sound.

The Sail Windmill

The sail windmill has its home in the Mediterranean countries, Crete in particular. In Crete alone they number tens of thousands and are used for water pumping. The sail mill is probably the simplest and safest mill for home construction. Despite its low tip speed ratio of 0.75:1 it has many advantages. Being made from wood and cloth it is inexpensive and easy to repair. Its slow speed and high rotor solidity make it responsive to low wind speeds. With a maximum speed of about 50 rpm, it is safe and moreover, at that sort of speed the white revolving sails set against a blue sky are a delight to watch.

Photograph 3. Sail wind generator, a 'soft technology' which generates up to 300 watts.

The sail mill is, as its name implies, made from sail cloth which is attached to a number (usually eight) of wooden spars. The rotor solidity ratio varies from 0.1:1 as in photograph 4 to 0.6. The lower ratio applies in high winds when the sails should be reefed (wrapped around the spars) to avoid tearing. The whole rotor structure is strengthened by tie-wires extending from a forward extension of the shaft, called a 'bowsprit'. Apart from its suitability for mechanical purposes, the sail cloth can be geared up to generate a small but useful electrical output. A speed increasing gear of between thirty and forty to one is needed if it is to drive an automobile generator. Its a cumbersome ratio, but can be done using gears and one or two stage belt drive.

Home construction of a sail mill can be very inexpensive, especially if second-hand auto parts are used – such parts are ideally suited for the shaft, brake drum, gears and alternator. The sail cloth and wooden spars are also inexpensive, but remember that where the cost in cash is low, the cost in time tends to be high.

The sail mill is self-governing to a degree in that it will spill excess winds by flapping and thus lose its natural aerofoil shape. The big disadvantage of the sail is that this self-governing action is of little use when the wind is excessively high. In such winds it is necessary that the rotor solidity be reduced, which in turn greatly reduces the force acting on the rotor and its support. Reducing the solidity means reefing the sails, which is pleasant work in the summer, but a totally different matter if you have to get up in the dark of a winter's night, stumble through the pouring rain to reach the mill, and struggle against the wind to reef the sails.

Actually, it sounds worse than it is, with a good knowledge of local wind speed, the above need rarely happen, or if it does, the damage done is usually just a few torn sails. Provided the demand for power is low, the sails may be left partially reefed throughout the winter season. As an alternative to reefing, a simple hook arrangement can be made for fixing and removing separate sails; leaving, for example, two instead of eight sails operating in high winds.

Whilst there is little written on the sail mill, compared to the high speed types, there are available four sets of reasonably good home construction plans detailed below, for the publishers see the bibliography.

31

Do-it-Yourself Sail Windmill Plan describes how to build the 12 foot diameter sail driven generator shown in photograph 3. It starts charging in an 8 mph wind and generates a useful 200 watts at 15 mph.

Reinforced Brickwork Windmill Tower gives, despite its title, plans for the construction of the 40 foot diameter mill and the brick tower shown in photograph 4. It is used for pumping water and starts operating in winds of 3-4 mph, producing a maximum of 13 horsepower (hp) in 25-30 mph winds.

Photograph 4. Forty foot sailcloth mill on a brick tower.

25 foot Diameter Sail Windmill is a design manual prepared by wind workers for the Brace Research Institute in the U.S.A. It produces power at 5 mph and a maximum of 6.7 hp at 20 mph. It has six instead of eight sails.

Food from Windmills describes in detail the construction and operation of 11 foot diameter sail water pumping mills built by the American Presbyterian Mission in Ethiopia.

The overall efficiency of a sail driven generator is at least 15 per cent and can be as high as 25 or 30 per cent, which is not bad considering the average 50 per cent efficiency of high speed propeller driven generators. I am not too concerned about theoretical 'efficiency' when it comes to wind generators. Surely, a wind machine is efficient if it suits the purpose for which it was designed and does not waste resources. It seems futile to expend great effort in designing a super efficient propeller when the same result can be obtained simply by increasing the propeller diameter.

Multi-blade Mills (or Fans)

There are few who have not seen one of these water pumping mills, for they still dot the countryside, some in action, but many in ruin. The nineteenth century saw hundreds of thousands of such mills in operation across the country, indeed there are companies who have continued to manufacture them from those days right up to the present.

The swept area of many multi-blade mills is more or less 'solid' with blades, in other words they have a high propeller solidity. Consequently, they are low speed mills with a tip speed ratio of less than 1:1, which is similar to the sail mill. Although the sail mill is regarded as more efficient, the multi-blade converts low wind speeds into useful torque power suitable for water pumping and other mechanical purposes.

33

Photograph 5. Multi-blade mill, now used to pump water through an array of solar collectors.

Photograph 6. A 25 inch diameter ventilation fan which is used to generate 50 watts in a 30 mph wind.

Four-arm Mills

This is the classic windmill of which few remain, and those which do are generally 'retired'. Some of these old mills were remarkably advanced in propeller design, efficiency and strength. At the beginning of the last century, propellers 100 feet in diameter were frequently used in Holland. Some of the old Dutch mills have recently been converted to generate electricity, one converted mill was originally built in 1727.

The Dutch four-arm has made a come-back at Santa Nella in California. There, a 46 foot diameter mill rotates on top of the 'Pea Soup Anderson's' Restaurant. In a 20 mph wind, frequent in that area, the mill generates a useful 8 kW of electric power. Designed by Wind Power Systems Inc., it is capable of withstanding gusts of up to 120 mph with the blades stopped. Automatic brakes hold the propeller in winds over 25 mph.

Photograph 7. This recently   built 'Dutch' four-arm generates a useful 8 kW at an overall efficiency of 20 per cent.

Photograph 8. The 'retired' four-arm still retains its grace.

Advanced Propeller Design

Most commercially available wind generators today use two or three long and slender blades. The design of the blades is such that they produce maximum lift and minimum drag. It is as a result of this 'lift' that the blades attain their exceptionally high tip speed ratio of between five and ten times the wind speed. This, together with efficiencies approaching 80 per cent, makes the two or three bladed propeller suited to the generation of electricity.

The very low solidity ratio of two bladed propellers give them a slightly higher tip speed ratio and aerodynamic efficiency than their three bladed counterparts. Twin blades are rarely used on wind generators with outputs in excess of 1 kW. The reason for this is vibration, which so frequently sets in. I know of one person who fixed a two bladed Winco 200 watt to his stone house, and even though the walls were thick, the vibration was transmitted from the tower through the

Photograph 9. With a maximum tip speed of 200 mph this little Winco 200 watt moves fast and furious, the blades do actually bow backwards in high winds, air flaps prevent overspeed.

37

walls to the shelves where the crockery and glassware inched their way to the edge. There seems to be two causes of the vibration, first that the effects of air tubulence may be increased by a bounce or shadow effect on one blade as it passes the tower, and second that if the main tower bearing is not rigid there tends to be a rocking of the whole wind generator.

Exceptions to the use of the twin bladed propellers, aside from small generators, are the huge Smith Putman and NASA wind machines with propellers of 175 and 125 feet in diameter respectively. The massive Smith Putman wind generator, which had a maximum output of 1250 kW lasted for several years, until one of its blades snapped at the root and that was the end of that. The NASA twin bladed 100 kW mill has not had a very impressive record since it was operational for only thirty-six hours. But yet, there are plans to build many more such turbines!!

For all practical purposes, the three bladed propeller is

Photograph 10. Three bladed direct drive Elektro, the spikes at the hub are speed govenors.

just as good as the twin bladed, and has the added advantages of greater stability and a lower starting speed due to increased blade solidity. Four bladed propellers, popular back in the thirties, have now been replaced by the three bladed version. Some twin bladed mills use two weights to give the stability of four blades. No matter what the blade configuration is, correct blade balance is essential for stability and each blade must be exactly balanced so that the weight is evenly distributed around the hub.

Blades are made from all sorts of materials — metal, wood, fibreglass, cloth and some reinforced plastics. Metal blades have been known to cause some radio and television interference, but other than that they tend to be reliable. By far the most popular are wooden blades which require particular care in their shaping if they are to give a good life of long service. The leading edge of each blade requires reinforcing with some extra hard material. Copper strip is often used for this purpose and is stapled into the wood, a practice I am not fond of since the staples can weaken the grain of the wood. I do know of at least one case where rot developed in the blade as a result of water seeping in alongside the staples. All blades need a protective coating of varnish or some hard wearing paint.

Cloth can be used as a blade or 'sailwing' when it is stretched as a sock over a metal and wire frame to form an aerofoil section. This type of propeller is suitable for home construction. Plans are published in the New Alchemists Journal Number 2. Much that is said in favour of the sail cloth mill can be said for the sailwing, but the sailwing has two great advantages. It has a much higher tip speed ratio and does not need reefing. In the event of extreme high winds, any damage which may result from flying sail material would be considerably less than that sustained by wood or metal. There is much to be said for the sailwing propeller. For home construction it is probably the ideal.

Photograph 11 18ft diameter sailwing at the New Alchemy Institute, used for water-pumping but can generate electricity if given a 25:1 gear up.

Vertical Axis Rotors

The chief advantage of the vertical axis rotor is that it does not need orientating into the wind, it is omnidirectional. A vertical axis Savonius rotor may be made by cutting an oil barrel in half and placing the two halves as in Figure 6. The Savonius rotor is slow speed, and with its efficiency of about 20 per cent, is mainly suited to water pumping, though there are a number of people who use the rotor to generate a low electrical output.

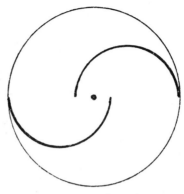

Figure 6. Plan of a Savonius rotor.

The Darrieus vertical axis rotor, and its many variations, have a far greater potential for the generation of electricity than the Savonius, in fact, there are three different Darrieus type wind generators available today. The Darrieus has speed and efficiency characteristics similar to that of high speed propellers as can be seen in the graph. All Darrieus

Photograph 12. Home built oil barrel rotor.

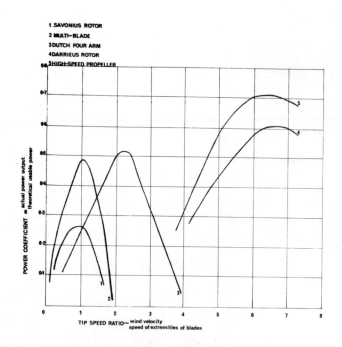

Figure 7. Tip speed ratio against power coefficient for various windmills. Note that each type of windmill operates most efficiently at a particular wind speed.

42

Photograph 13. Silhouette of a twin bladed Darrieus rotor.

rotors have two or more long, thin blades which may be curved (like a hoop) or straight, and which rotate around a vertical shaft. Originally developed back in the thirties, many vertical rotors still have a space age look about them.

The DAF rotor manufactured in Canada, produces 4 kW at 25 mph with a 15 foot rotor, or 6 kW with a 20 foot rotor, see figure 8. The rotor is not self-starting, it requires a push from the motor to get it going. With the DAF this happens every fifteen minutes. If the wind is strong enough the rotor will continue spinning (a most unusual sight) and if not it will stop. This seems a wasteful system which could be much improved by the introduction of a wind pressure switch which would start the rotor when the wind speed warranted it. Another disadvantage is that the guy wires which extend from the top of the shaft to the ground very much restrict the tower height, but then again this becomes a theoretical consideration of no importance if the rotor works successfully and at a reasonable price. The standard 8 foot tower

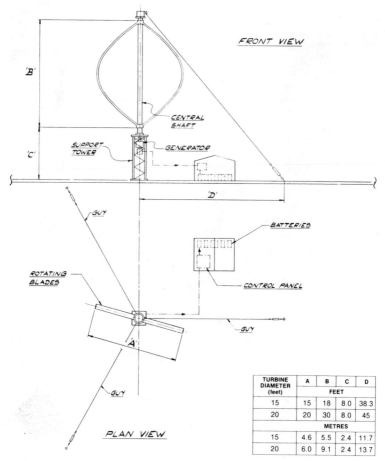

TURBINE DIAMETER (feet)	A	B	C	D
	FEET			
15	15	18	8.0	38.3
20	20	30	8.0	45
	METRES			
15	4.6	5.5	2.4	11.7
20	6.0	9.1	2.4	13.7

Figure 8. Vertical Axis Wind Turbine.

makes the generator accessible from the ground.

The Cycloturbine, developed by Pinson Energy Corporation, is a straight bladed Darrieus. As a result of its variable pitch blades it has the singular advantage of being self-starting. The tail or vane at the top of the rotor shaft, see photograph 14, senses the wind direction and adjusts the blades accordingly. Another interesting and recent development is the twin

Photograph 14. The variable pitch Cycloturbine, rated output 4 kW at 30 mph.

bladed variable pitch rotor shown in figure 9. The twin blades are hinged to the cross arm and are allowed to vary their pitch to suit changing windspeeds. This also avoids the extreme bending stresses caused by high rotational speeds. It is as a result of these stresses that the DAF rotor is curved, the curve (called a troposkien) is the same shape as that obtained by rotating a rope about a vertical axis. The expense of fab-

45

ricating the curve is avoided, using straight blades – unlike the Cycloturbine, the rotor shown in figure 9 is not self-starting. Neither of the variable pitch rotors require restricting guy wires as the DAF does.

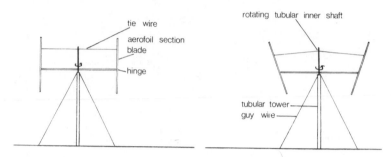

Figure 9. Variable pitch rotor configurations at low wind speed (left) and at higher wind speeds (right).

An arrangement similar to the propeller sailwing may be used on a vertical axis rotor, see figure 10. The sailwing

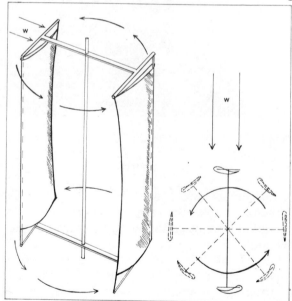

Figure 10. Vertical Axis Sailwing.

46

*Figure 11. Detail of vertical
axis rotor in photograph 15.*

*Photograph 15. Vertical axis rotors come in all shapes and sizes, above
is an Elektro on a mountain rescue post in the Swiss Alps.*

Photograph 16. An alternative twist to wind power, the Helical double rotor.

Figure 12. The Tetra-helix.

held in tension, achieves a variable pitch aerofoil profile as it rotates. Another recent development in wind power is the rotor shown in photograph 16. Each blade is twisted in its length through 180 degrees to form a helix. The rotor is self-starting. Although the tip speed is similar to wind speeds the rotors have a high aspect ratio of four (length/diameter) and therefore have a relatively high shaft speed. The prototype, shown in photograph 16, consists of two Helical rotors each 8 feet by 2 feet, and capable of generating 1 kW at 28 mph. Figure 12 is a similar rotor, which uses sail cloth, developed by Zephyr and called the Tetrahelix.

Photograph 17. Mr. A. Flettner's rotor driven boat.

In photograph 17 can be seen Mr A Flettner, standing by his rotor driven boat. It is said that he succeeded in crossing the Atlantic in a rotor driven ship in 1925! He also suggested that the rotors, which are cylinders spinning about their axis, could be used as blades for windmills. Flettner rotors operate upon the 'Magnus effect'. When a cylinder spins at a sufficiently high speed, the wind flow around the cylinder is unsymmetrical and a pressure is exerted on the cylinder in a direction perpendicular to the wind − and so the cylinder acts like an aerofoil.

Governors

To prevent propellers or rotors from overspeeding, governors must be used. A high speed wind machine without an efficient governor is a danger to life and limb. This is one of the main reasons why I do not encourage the home construction of such machines — unless one is going to put a considerable amount of time and effort into the work, and if that is the case why not go into limited manufacture, serving your own locality? The more local manufacturers the better, but please do not underestimate the considerable effort, time and study required to build a reliable wind generator. The same initial effort is required whether one or twenty-one machines are built.

To illustrate the problem of governing let us take an example of a twin bladed propeller, with a tip speed ratio of 7:1, which is geared up to a generator. Assume that this mill has been built from one of two sets of plans (one still available) neither of which bother to mention governors. At a wind speed of 25 mph all is well, power is generated and the propeller tip speed is only 175 mph. Should the wind suddenly increase to 60 mph before there is time to manually operate the flimsy hand brake, then the propeller tip speed will increase to over 400 mph. At such speeds any weakness or inbalance in any part of the machine will come under extreme pressure. Even the slightest inbalance in the propeller will invariably set up destructive vibration. That the gears or generator (or both) will probably disintegrate is of little importance compared to the potential hazard of blade shatter.

Fortunately all windmill manufacturers and careful home builders are acutely aware of this problem and build reliable speed governors into their designs. Of the many types of governor employed today, three are most popular. The first system uses weights attached or somehow connected to the blades. These weights, operating under centrifugal force, alter the pitch of the blades so that they lose much of their aerodynamic properties in excessive wind speeds and so maintain a reasonable speed. A second method is to place the propeller slightly off top-dead-centre of the tower shaft. The tail

Figure 13. The Winco airbrake in normal (left) position and governing (right).

vane is hinged so that in high winds the propeller is turned out of the wind, by wind pressure, as the tail folds into a plane parallel to the propeller. As the wind speed declines the tail unfolds, under force of gravity. The third system utilises air flaps or spoilers which have a drag effect and thus act as a braking mechanism. All such flaps, whether they are fixed on the hub or the blades themselves, operate under centrifugal force.

Wind machines have, or should have, manually operated brakes to facilitate servicing. Most manufacturers offer, as an extra, an automatic propeller brake which comes into operation in winds over, say, 60 mph. These brakes are usually operated by a wind pressure switch. As wind speeds in excess of 60 mph exhibit rapid changes in direction it is very wise to have the machine shut down in such conditions. An automatic brake will extend the life of a wind generator, particularly one sited in a high wind speed area.

Generators

All the old windmills, back in the thirties, used purpose built low speed DC generators. Many of these copper laden generators are still operative today. They had one drawback and that was the need for carbon brushes to collect the current. In some cases these brushes would last up to twenty years, and in others they might last only one year. To build such generators now would be prohibitively expensive, mainly due to the mass of copper required. Modern AC generators, properly called alternators, are lighter, smaller and therefore less expensive. Moreover alternators do not require brushes, even though they can be 'rectified' to produce a DC current.

The difficulty with most alternators is that they require a high rotational speed, at least 15,000 rpm. To take the propeller drive of 300-400 rpm maximum, from the shaft to the alternator, some form of speed increasing gear is required. Gears cost money, add weight to a wind generator and waste what would be otherwise useful energy; moreover, if they are not over-designed they may become the weak link in a wind system. In reaction to this a number of manufacturers have developed special low speed alternators suitable for direct drive from the propeller.

Towers

Earlier on in this book, under 'Site Selection', the benefits of increasing the tower height were pointed out. Here we will look at the various types of tower available. An example of the cheapest type of tower or mast is shown in photograph 18, where a telephone pole is used to support an old and heavy Jacobs mill. The 70 foot pole is tied with guy wires and hinged at the bottom to a concrete base. As a result of the hinge the whole tower, with the Jacobs on top, may be lowered and raised for servicing by means of a winch. It is also good to

Photograph18. Jacobs windmill supported by a telegraph pole.

know that should 'freak' winds occur the mill can 'lie low' in safety. The main difficulty with wooden poles is that the bit left underground eventually rots, but with this system that need never happen. Street lamp posts may be used as towers. It is of the utmost importance to understand the pressures acting upon the tower and to ensure that it is capable of taking such pressure.

Most people prefer to purchase purpose built towers from the wind generator manufacturer who understands what

53

Photograph19. Self-supporting tower with 3.5 kW direct drive Elektro in furled position.

is required in the way of propeller clearance, etc. Self-support-ing or guyed towers may be used. Self-supporting towers tend to be more expensive, but obviate the need for guy wires, which require regular and careful maintenance. The tower base should be well encased in a concrete cube, detailed instructions are supplied by manufacturers and differ with each type of tower.

Towers are assembled on the ground and then erected by means of a mobile crane or a gin pole. You will need friends around to assist with the gin pole method, figure 14, and the more the merrier — make a party of it. The tower base is set against a backstop, behind which is the gin pole. With bulkier self-supporting towers the gin pole may be set further back from the backstop. The gin pole should be a good height, at least 35% of the tower height. People are required to give initial lift but after that a car, or more people should finish the job. Once erected the tower is secured according to the manufacturer's specifications.

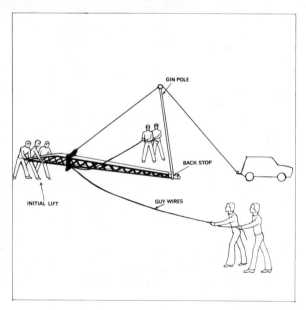

Figure 14. Raising a tower.

Photograph 20. The slender guyed tower is raised, with generator in place, by a crane; the propellers to be fixed when erected.

The next job is to fix the wind generator on top of the tower. This work is made easy with a crane, but aside from the expense, cranes cannot always go where wind generators are sited. In many cases, where the machine is not too heavy, it is possible to fix the windmill on the tower top before it is raised. This is always the case where a hinged tower is used. In other cases some manufacturers supply a small crane which is fixed to the top of the tower and used to haul the generator to the top, where it is swung in to place. Wind Power Digest (see bibliography) Number 5 gives details on how this is done.

The tower should be accurately vertical, otherwise output may be impaired due to a tendency for the wind generator to list a little out of the wind. All towers need lightning conductors and if there are guy wires it is advisable to earth these as well. Electric cables carrying current from the generator to the house should also be earthed, before they enter the house.

6 Electricity

Here we are going to be talking a lot about volts, amperes and watts, so let's get a thorough understanding of what these terms mean.

Voltage is a measure of the pressure under which electricity flows. The lower the pressure the slower the flow, any voltage (V) under 50 V is regarded as low. Car batteries operate under a pressure of 12 V, truck batteries are generally 24 V. Both voltages can give a mild shock if there is any human contact. Higher voltages, such as the British public voltage of 240 V can give a nasty shock, particularly if the skin surface is wet. Any contact with this should be avoided. No home wind energy system will ever need to use voltages higher than 240 V, which is just as well since contact could be fatal.

Whereas voltage is the pressure, amperes or amps indicate the flow of that electricity. Amperes can range from a fraction of an amp to the few hundred amps required to start a car motor. The greater the flow of amps through a wire the wider its diameter should be, much the same as water flowing through a pipe. The reason why most house fires occur is that an excess of current (amperes) is drawn through a wire too small to carry that load. The excess is frequently the result of multi-socket adaptors with, for example, two 25 amp heaters plugged into one 25 amp socket. The result of such action is that the wire overheats and may cause fire. It is to prevent this type of occurrence that fuses are used. Where an excess of current is drawn through a fuse, the fuse will simply melt. Replacing a blown fuse with an over-rated one defeats the purpose and causes house fires. The golden rule is never draw more current through a wire or cable than it is rated for.

Watts are a measure of the power used by an electrical appliance. Multiply the amps at which the appliance operates by the voltage and the wattage is known. So:

$$Watts = Amps \times Volts$$

from which follows

$$Amps = Watts / Volts$$

and

$$Volts = Watts / Amps$$

Hence a 12 V car light bulb which draws two amps is a 24 watt bulb. But take another 24 watt bulb, operating at a voltage of 120 V and it will only draw 0.2 amps. At 0.2 amps a much lighter wire may be used to carry the same 24 watts of power under the high pressure of 120 V as opposed to 12 V. Running four 100 watt light bulbs at 120 V would require only 3.3 amps, whereas the same wattage at 12 volts would draw a current of 33 amps, requiring a heavy wire.

If a 100 watt bulb is left burning for an hour it will use 100 watt-hours, and if it is left burning for 10 hours it will use one kilowatt-hour (1000 watt-hours equals one kilowatt-hour). Batteries are usually rated in amp-hours. Hence a 100 amp-hour battery will give one amp for 100 hours, or 100 amps for one hour. The amp-hour rating by itself is not very informative unless the battery voltage is known. In other words, a 100 amp-hour battery at *6 volts* will give only 600 watt-hours, whereas a 100 amp-hour battery rated at *120 volts* will give 12,000 watt-hours or 12 kilowatt-hours (kWh). A kWh is the standard electrical unit of measure.

There is one other aspect of electricity which requires

understanding and that is Ohm's Law. The ohm is a measure of resistance, by any material to the flow of electricity. Materials with very high resistance, such as plastic, are used as insulating materials. On the other hand metals, such as copper and aluminium, have very low resistance to the flow of electricity and are used as electric conductors. At low voltages, such as 12 V or 36 V, a loss of power can easily occur where a wind generator is sited far away from the house. Thick copper cable suitable for conducting 12 V current is expensive, and very expensive if one has to buy over one or two hundred feet of it. But what is possibly worse than the cost is the loss of power due to resistance in thick wire.

The resistance loss of copper and aluminium wire is shown below:

Wire Gauge A.W.G.*	Metric† Size	Resistance Ohms per 100' Two Wire	
		Copper	Aluminium
000	95 mm^2	.0124	.0202
00	70 mm^2	.0156	.0256
0	50 mm^2	.0196	.0322
2	35 mm^2	.0312	.0512
4	25 mm^2	.0498	.0816
6	16 mm^2	.079	.1296
8	10 mm^2	.1256	.206
10	6 mm^2	.1998	.328
12	4 mm^2	.3176	.522

* *American Wire Gauge (A.W.G.)*

† *The Metric system is now standard in the U.K., and replaces the Imperial Standard Wire Gauge. The resistances shown are exact for the A.W.G. sizes but only give a good indication of resistances found in Metric sizes.*

Table 6. Resistance losses of copper and aluminium wire.

All electric circuits have two wires, one positive and the other negative. The voltage drop in a wire is equal to amps times the resistance of the wire:

Voltage drop = Amps x Resistance

A drop in the voltage (due to resistance) will have an effect on the operation of most electrical equipment. The power loss in watts in a wire is equal to:

$$\text{Power loss (watts)} = \text{Amps}^2 \times \text{Resistance}$$

And if that has completely confused you I am not surprised. It was the same for me for a long time. However, the following example may help clarify the matter. Power generated by a wind generator is 600 watts. 200 feet of No. 4 AWG copper wire is used to carry power to the battery bank. Wire resistance, see above, equals 0.1 ohm ($0.0498 \times 2 = 0.1$). The choice of generator voltage is 12, 24, 32, 120 or 240.

Power generated 600 watts

Voltage	Output Amps	Line Voltage Drop	Voltage at load	Power Loss in wire	Power at load
12	50	5	7	250	350
24	25	2.5	21.5	62.5	537.5
32	18.75	1.87	30.12	35.15	564.85
120	5	.5	119.5	2.5	597.5
240	2.5	.25	239.75	.62	599.37

Table 7. The relation between generator voltage and power loss in wire with power generated by a wind generator of 600 watts.

In this particular example it can be seen that doubling the generator voltage results in one quarter of the power loss. There will always be some power loss in an electric line, but the object of understanding Ohm's Law is that the voltage and wire gauge may be chosen so as to minimise losses.

There are two different types of electricity, one is direct current (DC) and the other is alternating current (AC). AC current is supplied by the electricity board at a voltage of 240 V. This is the type of current and voltage which all domestic appliances operate on. Moreover current from the electricity board is supplied at a frequency of 50 Hertz. Any variation in this frequency causes appliances such as televisions and hi-fi equipment to 'brown' or 'black' out, but resistance heaters such as immersion or storage heaters will operate as normal on varying frequency and voltage.

Wind generators can produce DC or AC current. But because of everchanging wind patterns it is impossible to generate constant voltage and constant frequency electricity for use in the home. As such current from the generator is either AC, for heating, or DC for battery charging. In many cases the AC current from a wind-driven alternator is 'rectified' to become DC. AC current cannot be stored but DC can. The DC battery is an important link in a wind energy system. With a battery, power generated in high winds may be stored for use at any time. DC current from the battery can be used in a DC circuit, similar to that used in cars and caravans, or it can be changed to AC current, at any preselected voltage or frequency, by means of an inverter.

Batteries are expensive. Stationary batteries, as opposed to car batteries, are generally used with wind systems. Car batteries, new or second-hand certainly can be used, but they may be more trouble than they are worth. Car batteries are designed to give a high charge for a short period, whereas stationary batteries give a steady output over a prolonged period. Deep-cycle, lead acid stationary batteries have extra heavy plates and special separators. Life expectancy is 10 to 20 years with an efficiency of between 65 and 85%. Batteries become less efficient when they are cold and therefore may choose to keep their battery banks indoors. One word of warning on

Photograph 21. A battery bank capable of storing up to 12 kWh.

this, batteries give off hydrogen gas, and if this is allowed to build up, in a cupboard or wherever, it may reach a point where it will spontaneously explode. To prevent this the battery storage area should be well ventilated. It is the explosive nature of hydrogen that has hindered the development of the gas as a cheaper means of storage than batteries. It is possible to use a wind generator and, by a process called electrolysis, transform water into hydrogen and oxygen.

Stationary batteries are generally used for emergency lighting systems, golf carts, marine use, etc. Nickel cadmium batteries may also be used. New, they are very expensive but second-hand they may be a good buy. Efficiency is only 50 to 70% but they are suitable for total and high rates of discharge. Life expectancy is in excess of 20 years.

The size of the battery bank depends upon two factors, the local wind pattern and the allowable charging rate. The wind pattern is important as it establishes how long and frequent windless periods are over the year. In general four to seven days' energy needs should be met by the battery bank, more than that would be a costly extravagance. It is possible that a period of ten windless days will occur anywhere, but at such times it is more cost effective to cover this peak storage need with a stand-by generator. A characteristic of lead acid batteries is that they should not be charged too rapidly, or their life expectancy will suffer. Maximum charging rate is 14 amps per 100 ampere-hours storage, and this gives a rough way to assess the size of the battery stock required. In practical terms this means that even in areas where very favourable wind patterns would indicate the need for minimal battery stock, it nevertheless has to be big enough to take into account the charging capacity of the windmill. A voltage regulator is also required to stop the wind generator from overcharging the batteries, except in cases where the charging rate is less than 5 amps per 100 ampere-hours storage.

Inverters, which change DC current to AC, should be used as little as possible for they are both expensive and inefficient. There are two types of inverter – the rotary and the static. The rotary inverter is a DC motor driving an AC alternator. Efficiency is about 50 to 75%, but they also draw a 'no-load' current of 15 to 20%. A 1.5 kW rotary inverter will therefore waste up to 7.2 kWh a day, unless it is switched off when not in use. Rotary inverters are less expensive than static inverters, and when closely matched to the load there is not a great deal of difference between the two in inefficiency.

Static inverters are around 85% efficient and only draw a no-load current of 2 to 4%. Another important aspect of inverters is the wave shape they produce, which can be square or sine wave. Cheaper inverters produce a square wave. The importance of this is that most household equipment with motors are designed for operation on a sine wave as supplied by the grid. It is also important to ensure that any inverter chosen is capable of coping with the surge of power required to start motors.

To avoid the cost of an inverter, wherever possible use a DC system. DC motors are readily available, as are lights, in various voltages. The car and caravan trade use 12 V equipment. Many boats and trains still operate on 32 volts. To find

63

DC equipment can be a bit of a search, but it is worth it in the end. 12 V car bulbs last for years and years.

So, by use of batteries and an inverter, normal electricity board type power can be had from a wind generator. To add the convenience of the electricity board to a wind energy system a small back-up generator may be used to charge the batteries in times of low wind. The generator may be driven by petrol, diesel or propane. The output of the back-up generator can be a fraction of the rated capacity of the wind generator, which will provide essential power, or it can be equal to the full capacity. Whatever its capacity, a stand-by generator will only be called upon to supply less than three per cent of the total annual energy budget of the typical wind energy system. The generator can be manually or automatically started to charge the batteries when they are low. A wind energy system, complete with back-up generator, provides what is probably a more reliable service than the electricity board's. On the other hand, those who can go without electricity for a few days a year will get along fine without the additional generator.

A particularly interesting device, called the Gemini inverter, is now available for use with wind generators in association with a power line. Details of this system are given under 'Jacobs 2 kW Installation'.

WIND ENERGY

SYSTEMS

7 Owner-built

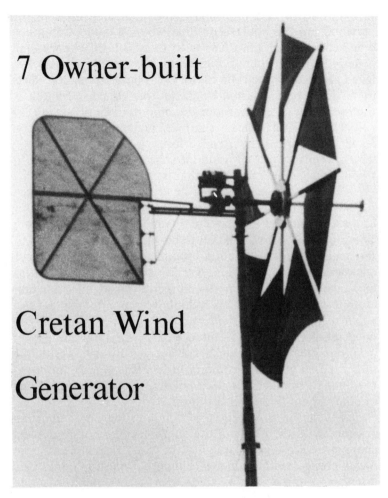

Cretan Wind

Generator

Photograph 22.

The 15 ft diameter sailcloth mill shown in photograph 22 generates 500 watts in an 11 mph wind. The maximum generator output is 1400 watts and the average daily output is 6 kWh. All the power generated is fed directly into two 700 watt 24 V DC immersion heaters.

67

The rotor hub is a sandwich of half inch thick marine plywood and this holds secure the two inch square Columbian pine spokes which in turn support the sailcloth. The windmill is designed to feather at a speed of 100 rpm. A specially designed gearbox with a ratio of 16:1 was built to take the speed up to 1500 rpm required by the motor vehicle alternator. At top speed the alternator will give an output of 60 amps at 24 volts (1.44 kW). The heavy shaft and manual brake came from a car scrap yard. It is intended to incorporate a safety device which will turn the tail vane into the same plane as the rotor when the wind gets too high, that is, over 30 mph.

The mast, bought second hand, is a 40 ft high lamppost purchased from the local council for a nominal sum. Its hinged base is encased in a concrete cube 3 ft in all dimensions. The hinge is to allow the mill to be winched up and down for servicing. At equal spacings around a 30 ft diameter circle anchorage points were dug. An inverted 'V' was made out of half inch steel rod and both end sections were buried in concrete leaving a small metal arch above ground. Stainless steel yacht rigging cable was used to secure the mast. Sailing equipment retailers can be a useful source for windmill builders.

The great, and single, disadvantage of this mill is that its sails need to be hand reefed in high winds. These are not so frequent in summer but it can be unpleasant winching down or climbing up the tower in winter.

The house in question is an interesting one, since it is designed with the whole of the south facing roof as an open channel trickle type solar collector. Water flows from the top down corrugated aluminium channels. This supplies summer hot water and contributes, albeit by a rather complicated heat pump process, to the annual winter space heating load. Added to this there is double glazing and a high degree of insulation. The floor is insulated with 2 inches of waterproof expanded polystyrene, the walls with 4 inches of fibreglass and the roof with 6 inches of fibreglass.

As well as being enjoyable places to be in, the garden and conservatory produce all the fruit and vegetables required by this family of four. A great deal of hard work went into making this house for the future a working reality.

8 Low Energy Winco

System

Photograph 23. Low energy cottage with solar collectors on roof.

The small four roomed cottage shown in photograph 23 is supplied with electricity from a Winco Windcharger rated at 200 watts in a 23 mph wind. The cottage lies, well protected from the wind, in a narrow valley whilst the Winco is sited 450 feet away on the crest of a hill. This is a difficulty, as transmission line losses are quite heavy on such a lengthy cable. It was to counteract this that a 24 volt model was selected − instead of the usual 12 volt generator. Even so, there is still quite a difference between power output at the generator and actual 'juice' at the other end feeding the batteries. With hindsight it can be said that a 36 volt model would have been a better choice. That aside, the Winco serves its purpose and lights the cottage.

The lighting arrangement is simple. 12 V DC car bulbs in

various wattages are used. Small 20 watt bulbs are sufficient for general lighting — casting, as they do, a soft warm light. Car bulbs can be arranged to give very pleasing indoor effects by using old headlamps to focus the beam. This way even a tiny 8 or 10 watt incandescent bulb will make a good bedside reading lamp. A 20 or 25 watt focussed bulb will give good light to a chess table, newspaper, guitar or whatever. A 35 to 50 watt bulb will give a clear, sharp light. This type of 'low energy' spotlight can be used just as well in the kitchen and eating areas as anywhere in the house.

It is true that fluorescent tubes are more efficient at converting electricity to light, but these tubes waste a lot of energy in lighting up the ceiling, walls and other useless places. They cannot be focussed. Hence I feel that the real difference in useful light intensity between fluorescent and incandescent is more or less nil, except in areas where a general diffuse light is required — such as stairs and corridors — there the fluorescent wins. One great advantage of the car bulbs is that they just don't seem to blow. I know of one family with a small

Photograph 24. Winco 200 watt high on a hill.

70

wind generator who, for the past nine years, have used car bulbs and, even though subjected to overvoltage, not one has blown in all that time. Finally, I think that the warm light from car bulbs (there are many different types) is much more friendly than the harsh fluorescent glare. However, enough of my biased opinions.

The air-brake on the Winco, composed of two flaps, can be seen in photograph 24. It prevents overspeed by 'spilling' the wind away from the propellers. One thing about this Winco which I found disquieting was the way in which the generator rocked on its main bearing in gusts of wind. This may have been due, in part, to a bad bearing, but it is more certainly the result of poor siting. The Winco appears to be on an ideal site — right in the middle of a tiny plateau on top of a steep hill. This is potentially an excellent site, but not when the standard 10 foot tower is the only support used for the Winco. What happens is that the wind rolls up the hill, and is then tossed in turbulence over the hill-top, buffeting the wind mill at the same time. To avoid this destructive action the tower should be raised by at least 20-30 feet. As I have mentioned before, a windmill operates best in a smooth airflow.

The cottage, which houses two, is heated by a wood-burning Jotul stove with maximum heat output of 6 kW. Solar panles on the roof supply all the hot water required in the summer and bottled gas supplies cooking and winter water heating. With a slightly larger wind generator rated at 500 to 1000 watts a significant contribution could be made to winter water heating. In such a case current for water heating would be fed direct to an immersion heater and thus by-pass the batteries. The cottage can be seen at the Centre for Alternative Technology, Machynlleth, Wales.

9 Jacobs 2 kW installation

The installation pictured here is located at an altitude of 1700 ft on a shoulder of Prickly Mountain in Northern Vermont, USA. It provides power for the home of Donald Mayer of North Wind Power Company (see Access), and is located in a neighbourhood that supports considerable experimentation in the field of alternative energy. Two other wind plants dominate the approach road. Three nearby homes utilise solar

72

heating systems. The tower, 60 feet tall, is located some 30 ft above and behind the house, on a granite outcropping. It is bolted into the ledge. The generator is a re-built Jacobs direct drive model, giving an output of 2 kW (110 volts) at a wind-speed of 20 mph. The blades were made by North Wind of a aircraft quality sitka spruce, finished with a thin layer of fibreglass and painted in white epoxy. The aerofoil design is an improvement upon that used by Jacobs.

The cost of the generator, governor and blades was £2000. The tower, a self-supporting model, cost £325, and the Gemini Synchronous Inverter cost £800. Installation, cal-culated at 12 man-days at £30 per day, cost £360. The total system, including miscellaneous parts, cost £3485. Due to the tower's height, it was necessary to obtain building permission. This created no problems.

The wind-plant provides 200 to 300 kWh per month, or about half the household's electrical consumption. In times of high wind speed and low electrical consumption, excess out-put is stored in the grid via a Gemini Synchronous Inverter, a solid state unit which forms the interface between the grid, the generator and the house wiring. When the wind fails, current is provided from the grid to the house through a standard electricity meter in the usual way.

The system is similar to that used for many years in re-generative drives such as lifts, decelerating electric trains and other heavy equipment. Its advantage is twofold. The unit converts the generator's variable DC output to constant volt-age AC, and synchronises the signal with the grid's sine wave. In addition, the wind system's overall cost is cut in half by elimination of battery storage and conventional inverter. In this case, the installation has been extensively monitored by the local electricity company so as to enable them to deter-mine an equitable rate structure for customers who do not utilise the power source as a primary electrical source.

There is difficulty in getting permission from the CEGB in the UK to allow this practice. As with any bureaucracy, it is frequently a matter of approach. If one puts forward a pro-posal to use such an inverter, and are half expecting them to

Photograph 26. 2 kW Jacobs Wind Electric.

put down the idea, then that is what will probably happen. On the other hand, if you go and meet the 'faceless ones' and be positive and enthusiastic about the idea as an experiment, you might be surprised by an agreement. The use of Gemini inverters is legal in France and in most US states.

The Mayer family had several reasons for wanting to utilise wind electricity as a source of energy for their home. As an expression of their philosophy of decentralization, the system retains the option of becoming fully independent at some future date with the addition of battery storage. The system increases awareness of energy consumption through visual impact. As such, it is also an educational tool in the local community, demonstrating wind power's validity as an energy source.

The other energy systems in the house are compatible with this philosophy. Cooking takes place on an antique wood-fired stove, with a modern propane stove serving as a back-up in hot weather. A Clivus Multrum composting toilet receives all human waste and kitchen scraps. Requiring no external energy input, the Clivus will eventually produce a high quality garden fertiliser. Through use of a Clivus, the 3 Mayer children have learnt to identify bio-degradable materials in their environment.

Space heating is a major use of energy in the climate of northern Vermont, and the Mayor home utilises passive solar heating techniques to provide for a substantial portion of the heating needs. Virtually the entire south face of the structure consists of a 45° pitched roof glazed with acrylic sheeting. The floor of the public living spaces consists of a heavy concrete slab faced with slate. During periods of direct or indirect solar radiation, the floor acts as a heat sink, aided in its storage capabilities by a massive central chimney. When the house is complete, panels of insulating material will be able to slide into place behind the glazing, retarding re-radiation of stored heat from the slab into the night sky. The insulating panels will be stored behind a bank of liquid-cooled solar collectors that provide most of the family's hot water needs. Supplementary heating during cloudy weather is provided by a space heater fueled with hardwood culled from the surrounding forest. A propane central furnace allows the family to be absent for several days at a time without fear of frozen plumbing. Overall heat requirements are reduced by the inclusion of three inches of urethane foam insulation in the building's exterior walls.

A clearing just in front of the house allows space for several beehives, a vegetable garden, and forage for two work-horses. These elements complement the Mayer's decentralist philosophy: developing the skills necessary for achieving a high degree of material self-sufficiency.

10 Hydrowind

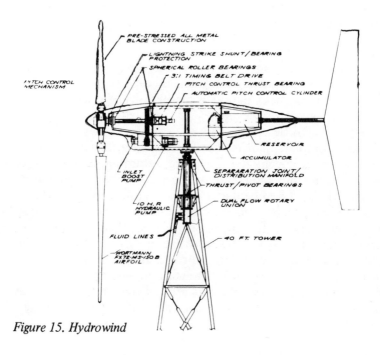

Figure 15. Hydrowind

The Hydrowind, erected in September 1976, was the first wind generator to utilise a hydraulic system. It was designed by Merrill Hall and Vince Dempsey, for use by the New Alchemy Institute at their Ark, a research centre on Prince Edward Island, Canada. P.E.I. leads North America on its laudable decision not to use nuclear energy, but instead to take a more positive role in utilising a coal/solar/wind energy system.

The New Alchemy Institute, a non-profit making organisation, was granted funds to build an 'Ark' on a coastal site of 137 acres. The Ark is an integrated structure, comprising of a residence, extensive greenhouse, fish farming unit, research laboratory, barn and tool shed. The south facade of the building is pure solar with a vast expanse of solar collector and greenhouse glazing. Rocks and water in the building store up to 13.2 million BTU's of solar heat which, when combined with a woodstove, will heat the Ark for all December, even if the sun never shines! From the residence, there is a spectacular view of the sea and the afternoon sun shines freely into the living room, dining room and bedrooms.

The Hydrowind, the first of four such machines to be used on this site, generates a maximum of 7 kW in a 25 mph wind speed. The three bladed upwind propeller has a diameter of 20 feet. The blades utilise a light weight design based on an internal tension system and are covered with an aluminium skin. Hydraulic governors are used to vary the pitch of the blades, thus giving a constant propeller speed in varying winds. A 3:1 belt drive connects the shaft to the hydraulic transmission, which in turn drives the alternator on a platform half way up the self-supporting tower.

The overall efficiency of the hydraulic system is about 90%. It is interesting that another company in England, (see the 'Elteeco System'), was working simultaneously on a very similar system, each unaware of the activities of the other. A Gemini synchronous inverter is used with the local grid to provide alternating current at 110 volts, 60 Hz.

The Hydrowind, still under test, is not yet commercially available, although the design is such that it is suitable for regional manufacture. The performance of the energy systems

*Photograph 27. The Ark under construction with New Alchemist
John Todd in the foreground.*

in the Ark during the severe winter of 76/77, was even better
than expected, which is very encouraging for them. The size
of the Ark probably represents the minimum climatic mass required for its operation as an economic unit with part-time
tending. It has the capacity to raise up to 20,000 fish at one
time, 10,000 cuttings of valuable fruit and nut trees, plus the
ability to raise year round greenhouse food and flowers within its integrated design. In building the Ark, the goal of the
New Alchemists is to show how to build living structures
which pay for themselves.

The excellent Journal of the New Alchemists and membership of the Institute is available from:

P.O.Box 432, Woods Hole,
Massachusetts 02543, USA.

11 The Elteeco System

This novel system, developed in England, has many attractions: it has a simple and sturdy blade configuration, it uses hydraulic gears and transmission. There are two ground level generators, one an induction generator for operation with a local power line to provide mains type electricity and the second, an alternator for resistance heating. Lastly, it is aesthetically pleasing, reminiscent of old-fashioned mills.

The Elteeco wind generator shown in photograph 28, the first of its kind, is sited near the home of the inventor, Sir Henry Lawson-Tancred. Rated at 30 kW in a 20 mph wind, it is principally designed to heat and light large country houses, although it can be used for many other purposes, such as greenhouse heating, etc.

The Elteeco uses a three-bladed, fixed pitch propeller made of steel spars supporting fibre glass envelopes moulded in an aerodynamic profile. The blade structure is strengthened in a basic and elegant way to withstand wind speeds of up to 120 mph. Guy wires connecting the centre of each blade can be seen in the photograph, as can the metal supports which connect to a central tripod.

The 60 foot diameter propeller is mounted on a horizontal shaft, which in turn is supported by a 45 foot high, four post, timber clad tower. The propeller shaft is not strictly horizontal, but is set at an angle of 5° so that additional clearance is provided between the blades and the tower base. The tail vane directs the blades into the wind.

Photograph 28. The Elteeco wind system. All the expertise, including manufacture of the 60ft propeller came from within a few miles of the windmill site.

The shaft is connected to four hydraulic gear pumps by a speed-increasing gearbox. As a result oil, under pressure, is passed to the 'energy integrator'. The energy integrator supplies hydraulic oil under constant pressure, through shut-off control valves, to two hydraulic motors of low and high power respectively. The low power motor operates a 5 kW induction generator in phase and voltage alignment with the mains supply from which it is energised. The high powered motor drives a 25 kW alternator, separate from the power line, and whose output is used for heating purposes only. Full output of 30 kW is reached in a 20 mph wind. The average wind speed on this, the test site, is 15 mph. The expected output per year is 90,000 kWh, 15% of which will be grid-type electricity from the induction generator, the remainder, from the alternator, will go to space heating.

Excess energy in higher wind speeds up to 27 mph is dissipated through a pressure-reducing valve. A further rise in wind speed causes centrifugally operated blade flaps to extend, thus providing an air brake on the propeller. The Elteeco System is equipped with a wind speed sensor, connected to an electronic circuit. In the event of wind speeds above 35-40 mph, spring loaded (but electrically retained) band brakes stop the propeller. There are further safety devices attached to both generators. In the past few months of testing, it has withstood winds of up to 90 mph.

Elteeco offer three wind systems; the first as outlined above, which is still under test; the second drives a 30 kW induction generator in parallel with a local power line, assuming the electricity board is agreeable to the arrangement; the third drives an ungoverned alternator suitable for heating purposes only. This produces 50 kW in a 25 mph wind and at the time of writing is available for £12,000.

It will be interesting to see how the Elteeco Wind Machine proves itself over the next year or two. Those seeking further information should send £1.00 to: Elteeco Limited, Aldborough Manor, Boroughbridge, Yorkshire.

12 Zephyr Wind

Installation

The house shown in photograph 29 was designed as an all electric house and that was back in 1970 when electricity was cheap. Within the space of a few years the cost of electricity had tripled. As there seemed to be no end to these increases, the owner of the house decided to invest in an inflation free wind generator. Today, with the help of two wood stoves, the Zephyr Wind Dynamo supplies space and water heating. Utility electricity is relegated to lighting, motors and back-up.

The Zephyr is a three-bladed downwind model which gives a maximum output of 15 kW in a 30 mph wind. The expected output on this site, a rocky shorefront with an average windspeed of about 15 mph, is about 20,000 kWh yearly. The electrical output has a varying voltage and frequency. As all the output is used for resistance heating, batteries and inverter are unnecessary, although these components can be added on later if an independent source of utility type electricity is required.

The Zephyr is wired to supply two separate 3-phase circuits of 7.5 kW each. Careful resistance matching is required to achieve best efficiency. This system uses four appliances which can be run selectively, First, a space heater with nine 1 kW incandescent bulbs; second a 7.6 kW storage heater made up from seventy building bricks with nichrome wires threaded through the cores; a waterbox located in the master bedroom is the third appliance. It heats the room as well as the incoming cold water before it enters the hot water boiler. It is valved

Photograph 29. Zephyr 15 kW wind dynamo.

so that in windy hours water may by-pass the commercial electric heater for direct use. The fourth appliance is a 15 kW immersion heater, used only in the summer for heating the swimming pool. A portion of the additional heat goes to a south-facing 200 sq ft growing room which is maintained at 50° at night. This glass-walled room accounts for more than 500 kWh per month of electricity in mid winter.

The total cost of duplicating this installation would be approximately £13,500. Expensive, and justifiable only on the grounds that fuel prices will continue to increase as they have over the past few years. The great benefit of this resistance heating system is that it involves no batteries or inverter. That just leaves the difficult question of how long the Zephyr wind generator will last. I cannot comment on the plastic composite blades, except to say that with a good protective coating I hope that they would last for many a long year. The direct drive feature means that there will be no future gear-box problems. The six-pole, permanent magnet alternator is of solid, low speed design and is well protected from vibration by rubber mountings. The bearings will need replacing once every seven years or so, but they are very cheap.

The owners of this installation, Giff and Lois Horton are skilled and energetic people, having, with the help of their sons, designed and built an innovative and energy conserving house long before it became fashionable. They are excellent customers for wind generators, and in particular for field testing new machines like the Zephyr, as they recognise the experimental nature of the design, and are prepared to deal with the many annoyances and difficulties inherent in the testing of a new product. This ability to grapple head-on with the technical, financial, legal and institutional problems raised by the introduction of a new natural energy device makes the Hortons and others like them the 'new pioneers'. For further details of the Zephyr wind dynamo see under Access.

13 Conservation House

Photograph 30. Conservation house is sheltered by trees, but so is the Dunlite which should be on a higher tower for better performance.

The house shown in photograph 30 is possibly the most advanced of its kind. It depends for its energy on a 2 kW Dunlite wind generator which provides about 15 kWh of electricity a day. This meagre energy input, approximately one fifth of the average home use, is capable of providing all normal domestic requirements — temperature control, cooking, water heating, lighting and small power.

The three bedroomed house is well wrapped in 18 inch thick fibreglass cavity insulation. This gives a 'U' value of 0.0737 and effectively reduces the energy required for winter

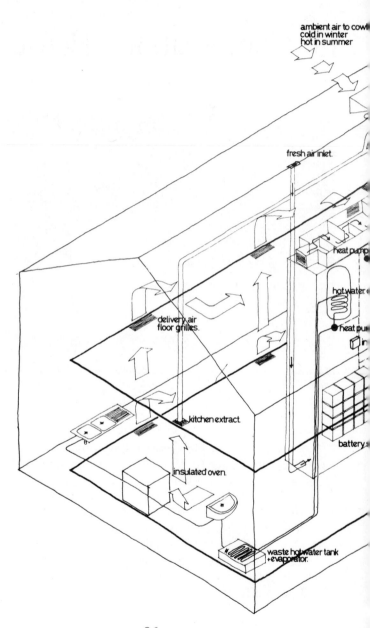

ambient air to cowl
cold in winter
hot in summer

fresh air inlet.

heat pump

hot water

heat pump
in

delivery air
floor grilles.

kitchen extract.

battery s

insulated oven.

waste hot water tank
+evaporator.

86

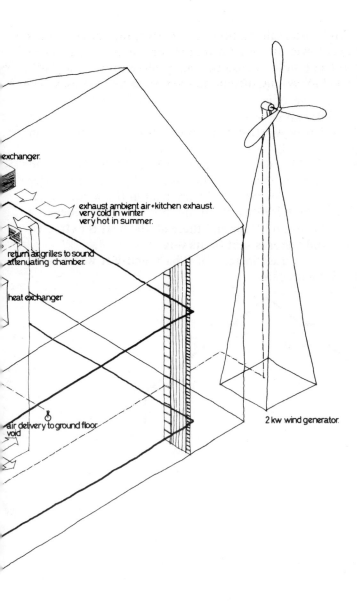

exchanger.

exhaust ambient air + kitchen exhaust.
very cold in winter
very hot in summer.

return air grilles to sound
attenuating chamber.

heat exchanger

air delivery to ground floor
void

2 kw wind generator.

*Figure 16. Air distribution system, hot water supply and electrical
layout for Conservation House at NCAT.*

heating to one eighth of normal. The windows are not just double glazed but have four sheets of glass! Temperature control is effected by a small heat pump driven by a 150 watt motor. The heat pump, which can be used to heat or cool, has an optimum co-efficient of performance of 3:1, which means that for every unit of energy input there is an output of three units of heat (see Air Distribution diagram). Air flow within the house is carefully controlled to avoid heat loss.

Domestic hot water is again provided for by a small heat pump, this time with a C.O.P. of 2:1. The heat for water heating is ingeniously extracted from heat in the waste water. Instead of pouring it down the drain, hot water first goes to a tank under the ground floor and there all the heat is extracted before the cold water is sent on its way.

A shower is used instead of a bath and hot water is, in general, used with care.

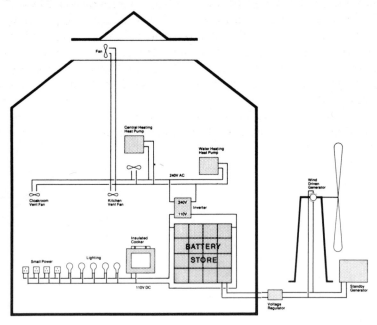

Figure 17. Electrical layout diagram for Conservation House at NCAT.

88

An electric cooker, operating on 110 volts DC, does all the cooking on approximately half the energy normally required. This is effected by means of a patented design which retains the heat in the oven; also the oven is enclosed in six inches of insulation. Lighting is provided by low wattage fluorescent fittings. Power socket outlets are provided for domestic appliances.

Power from the wind generator is fed to a 182 ampere hour, 110 volt battery bank. A small standby generator is available for any long windless periods and also to allow for periodic maintenance on the Dunlite.

Rain water is collected, filtered and used for all domestic purposes except drinking. Conservation House was designed for a site at latitude 52° but its principles could apply to a much wider spectrum of climates.

Details aside, and much as I appreciate the house, I do not feel that I would be happy living there. Once inside the house I find that it reminds me of an isolation ward in that one is cut off from the outside world — bird song or wind in the trees cannot be heard through the layers of insulation, moreover the small windows do not open. However not every one agrees with my criticism and I am glad that the house has been built.

Further details about this house are available for 50p the Centre for Alternative Technology, Machynlleth, Powys, Wales.

14 Wind Powered Solar Greenhouse

As the demand for oil exceeds supply, in a few years time we are going to have to reassess not only our domestic energy requirements but also our food requirements. I fear that the oil intensive agri-business which currently provides our food has a short life. A close look at modern agriculture and our food distribution network will show how totally oil-dependent it is. Anybody who does consider this dependence will, almost certainly, want to provide food for his or her own family.

The system shown in photograph 31, a wind generator and greenhouse, is an interesting one. When the wind blows, greenhouses lose their heat. The problem can be solved by insulating the greenhouse, especially the north wall and over night, and then adding a simple wind energy system to produce heat, replacing that which is lost simultaneously. In such a system, there is no need for batteries or inverter as all the current is simply fed into an immersion heater, similar to a kettle element, in a water tank or some other form of heating.

The wind generator shown in photograph 31 is not really essential because the greenhouse, originally developed by Helion, is designed efficiently and requires little external energy Double glazing and insulated walls help to retain heat during cold periods. A curved rear (north) wall acts as a light reflector to provide increased light to the interior during the winter months with low sun angles and short days. Plans for the home constructor who wishes to build this greenhouse are available

Photograph 31. Wind powered solar greenhouse.

from Provider Greenhouse, Box 49708, Los Angeles, California 90049, USA. This, however, is an exceptional greenhouse of recent design and does not help to solve the problems of existing greenhouses. Insulation will help a great deal, but additional power must be provided. In many cases, wind power, with or without a heat pump system, will be sufficient to maintain the right conditions. Wind Energy Supply Company (see Resources) has done considerable pioneering work in this direction. In particular, WESCO have developed a 60 ft diameter 100 kW wind turbine, intended for use with commercial greenhouses.

The smaller domestic system in photograph 31 features the Helion 12 ft diameter downwind turbine. Built by the owner from the Helion 12/16 Construction Plans (see Resources), it provides about 1500 kWh of heat per annum on this particular site with an average wind speed of 10 mph. The total cost of the wind generator was £650 and its construction took about ten man days of concentrated though enjoyable work.

15 Owner-built Multi-blade Windmill

Arnold Stead is a retired chimney sweep who enjoys playing the double bass and watching the stars through his telescope. The small house where he and his wife live is heated by a wood stove and the wood for this stove is cut by a circular saw driven by a gaily painted multi-blade windmill. The mill has worked happily and without trouble for the past fifteen years.

About twenty years ago Arnold wondered if he could put some of the waste he saw around the countryside to a useful purpose. Even though he had no previous mechanical training or experience, he decided to build a windmill. The blades and shaft of the first mill he built were not strong enough. Vibration set in and a 60 mph wind blew it to pieces.

Learning from this experience, Arnold set about building another improved mill and the results of his efforts can be seen in photograph 32. The 8½ ft diameter multi-vane propeller, which also has eight small blades at the hub, spins with the slightest breeze and looks like a colourful, revolving mandala. The propeller alone weighs a massive 225 lbs, and is built from scrap like everything else on this mill — old bed parts, second hand pulleys, even old railway tracks are used to carry the weight of the windmill on top of the workshop. If this sounds all very slapdash, it isn't, and it all fits and works. At the end of the propeller shaft is a 168 lb flywheel, again scrap, and it took Arnold six months before he found the right one.

Photograph 32. Arnold Stead's Pennine wind engine.

The main bearing is built to withstand the pressure of 5 tons. A crown wheel and pinion takes the drive straight down into the workshop, where it operates two grindstones, for sharpening tools and knives, and the saw. Exact performance figures are not available but the mill does its job, producing up to 5 HP (3.8 kW) in a 30 mph wind. It is capable of sawing fair sized logs. Moreover, it has stood the test of time, and windspeeds of 90 to 100 mph!

The secret of Arnold's success is the way in which he built the mill. It has worked well for fifteen years and looks fit to last at least another fifteen. Every night, before building the mill, Arnold would go to bed and instead of going to sleep he would shut his eyes and work the mill through to its finish. He visualised the mill, complete and working, before starting to build it. Then, knowing exactly what parts he wanted, he would go about his business as usual but keeping an eye out for the necessary ingredients. If this meant waiting for months, well, he could wait.

This reminds me of the millwrights of old who used to build the Dutch four arm mills. When visiting the site of a proposed windmill for the first time the millwright would invariably infuriate his employer due to his habit of standing for hours and even days just looking at the site and apparently not doing any work at all. Finally, and in his own good time, the millwright would start the construction work. He had probably spent this time of physical inactivity looking for the best site and visualising how he would build the mill.

I am sure that the more creative and careful thought put into a system or machine before its construction, the longer its life will be and the less trouble it will give. The rapid obsolescence and shoddiness of much of our industrial produce is witness to the lack of careful consideration built into the original idea. On the other hand, Arnold Stead's mill is endowed with an almost life-like quality.

ACCESS

Manufacturers and Rebuilders

Aero Power,
2398 4th Street, Berkeley, California 94710, USA

Aero Power have been in the wind business for about six years and during that time they have manufactured about 700 sets of propeller blades for a wide variety of reconditioned machines and for colleges.

Aero Power also manufactured a 'Model A' twin-bladed wind generator rated at 1000 watts in a 32 mph wind. This has now been replaced by a new three-bladed 'Model A' rated at 1000 watts at 25 mph.

The propeller has a diameter of 8 ft 6 in. The variable pitch blades are made of Sitka Spruce. Transmission to the alternator is by means of a hellical gear with a ratio of 2.5:1; The alternator starts charging at 42 watts in a 7 mph wind and reaches a maximum output of 1050 watts at 25 mph. At high wind speeds the blades automatically feather preventing damage to the unit.

The price of the Aero Power with control box is £1060. They also sell Rohn towers, inverters and batteries. Send £0.50 for their catalogue.

The following is an example taken from their range:

60 ft guyed tower	£ 345
500 watt square wave inverter	£ 200
12 volt 230 amp hour battery set	£ 148
Aero Power wind generator	£1060
Total	£1753

American Wind Energy Association,
Box 329, Route 3, Mukwonago, Wisconsin, USA.

The AWEA is a national association which represents manufacturers, distributors, and researchers involved in the development of wind energy. Membership is 25 dollars per year.
The AWEA issues a quarterly newsletter, publishes the 'Wind
Technology Journal' and holds convivial and enlightening con-
ferences.

Aerowatt,
37, Rue Chanzy, 75011 Paris, France.

Aerowatt manufacture the most expensive range of wind generators available. One reason for the expense is that their
machines are rated at very low wind speeds. For example,
their 4 kW mill gives its full output in a 15 mph wind and will
cost a mere 77,000 French Francs, their 1 kW mill costs
35,200 Francs. The other reason for their cost is that they are
built to operate in very severe windspeeds of up to 175 mph.

American Energy Alternatives (Amernalt)
Box 905, Boulder, Colorado 80302, USA.

Amernalt currently offer two wind generators. The 8 ft diameter horizontal axis rotor is similar to the old multi-blade
mill. One machine is rated at 1.5 kW in a 26 mph wind and
costs £1975, the second rated at 2.5 kW at 40 mph costs
£2080. A 12 ft diameter rotor will be available shortly, generating 2.5 kW at 24 mph. Both machines have withstood
winds in excess of 90 mph. The aluminium blades are fixed to
steel spokes. The unit is fitted with an automatic brake, and
comes complete with electrical overload protection, voltage
regulator etc.

Photograph 33. An Aerowatt mill.

American Wind Turbines,
1016 East Airport Road, Stillwater, Oklahoma 74074, USA.

Manufacture a multi-blade rotor similar to the Amernalt. Their 16 ft wind generator reaches its maximum output of 2 kW at 20 mph (1 kW at 15 mph). Their prices are put tentatively as follows:

Wind turbines	£ 525
16 ft tower	£ 430
Mounting kit	£ 75
Alternator and controls	£ 360
Total	£1390

Photograph 34. The American Wind Turbines.

AWT also manufacture 12 and 16 ft turbines suitable for heating, powering electric motors or electric water pumps, with prices similar to those above. The use of a switch box (£40) will allow one turbine to serve all three purposes.

Ampair Products,
Aston House, Blackheat, Guildford, Surrey.

Manufacture 50 watt wind generators for marine applications.

Battery Manufacturers

The cost of batteries varies greatly. A small local manufacturer may charge considerably less than a national producer. Occasionally good quality second-hand batteries may be purchased. Consult Yellow Pages for details of manufacturers.

Robert Bosch GmbH,
33, Osborne Road, Southsea, Portsmouth, Hampshire.

Manufacture a range of capacitors which can be useful for starting motors with a battery/inverter system.

C.F.Casella and Co Ltd.,
Regent House, Britannia Walk, London N.1.

Manufacture a range of anemometers.

Compcop,
Box 1267, Redwood City, California 94064, USA.

Compcop sell a wind generator kit which costs £150 less the generator. The three bladed 12 ft diameter aluminium propeller generates a maximum of 500 watts in a 20 mph wind when fitted with a suitable car alternator. The plans cost £4 and a catalogue £1.

Coulson Wind Electric,
RFD-1-Box 225, Polk City, Iowa 50226, USA.

Roland Coulson, 'the answer is blowing in the wind', sells a wide range of re-conditioned wind generators originally manufactured back in the thirties and forties, such as the Air Electric, Delco, Parris-Dunn, Wincharger and Windpower.

Air Electric Company produced two 32 volt machines, 2 kW and 3 kW. On both models, the generator cowling and tail were one continuous piece with vents for air cooling the generator. The two bladed propeller had an 80 lb flywheel for

Photograph 35. Windpower 2.5 kW built in 1934.

stability. Governing was achieved by means of curious looking paddle deflectors.

Delco and Parris-Dunn both manufactured small 200 watt wind chargers for radio operation. In high winds, the Delco tail vane would fold and thus draw the propeller out of the wind. Parris-Dunn, who also manufactured a 2 kW model, had

WIND—H

103

an unusual governing system in that high winds would lift the propeller upwards into a vertical plane out of the wind.

Wincharger Corporation started in 1927 and manufactured a few hundred thousand wind generators before being bought by Dyna Technology Inc., who continue to produce the Winco 200 watt. Wincharger also produced two and four bladed 650 watt and 1200 watt models. The two bladed machines all used centrifugal air flap governors, as does the Winco 200 watt today.

Windpower Corporation produced 1.2 and 1.8 kW three bladed downwind machines. The 12 ft diameter propeller was directly connected to the 32 volt generator. The blades were governed by flyballs. A manual shut-down brake was used in storm conditions.

Coulson also sells re-conditioned Allied, Black Swan and Alamo Dynamo wind machines; new and used batteries; towers and 32 voly equipment.

Coulson powers his workshop with a 1.2 kW Wincharger, on an 80 ft tower and a 2.5 kW Windpower, on a 65 ft tower. All the workshop equipment is run on 32 volts DC, lathe, valve and bench grinders, drills, air compressor, vacuum cleaner, radio and lights. Coulson's is an emporium for re-conditioned machines and parts.

Delatron Systems Corporation,
553 Lively Boulevard, Elk Grove Village, Illinois 60007, USA.

Recently announced the availability of a new range of cost competitive DC to AC inverters developed for wind and water power users. There are three initial models:

36 V DC to 120 V AC, 3 kW capacity	£1600
120 V DC to 120 V AC, 3 kW capacity	£1600
120 V DC to 120 V AC, 6 kW capacity	£2285

This looks good but write to them for more information. Delatron also manufacture heavy duty batteries.

Dominion Aluminium Fabricating Limited,
3570 Hawkerstone Road, Mississauga, Ontario L5C 2V8,
Canada.

DAF manufacture vertical axis Darrius turbines. Both the 15
and 20 ft models require motor start. The 15 ft model is rated
at 4 kW in a 23 mph wind, the 20 ft model generates 6 kW at
the same windspeed. The choice of generator voltage on the
15 ft model is 24 or 110 V. It is designed to withstand gusts
of up to 130 mph. The 15 ft unit comes complete with electrical control gear and stub tower, and costs about £4800.

Dunlite
Pye Industries Sales Pty Ltd., 22 Hargreaves Street,
Huntingdale, Victoria 3166, Australia.

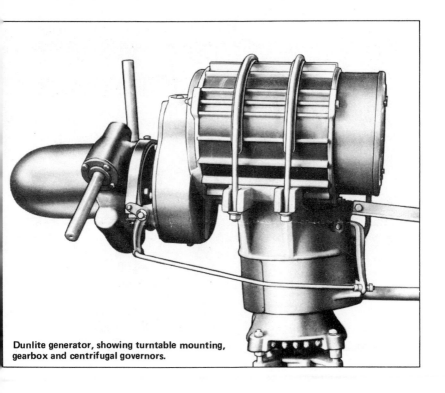

**Dunlite generator, showing turntable mounting,
gearbox and centrifugal governors.**

The Dunlite 2 kW wind generator has been on the market for 30 years and in that time has proved itself to be a sturdy and reliable machine. Three galvanised steel, variable pitch blades form the standard 13 ft diameter propeller. Centrifugal gov-

ernor weights are mounted on the blades. A heavy duty helical gear takes the maximum propeller speed of 150 rpm up to
the alternator's 750 rpm. The specially built multi-pole brushless alternator generates a maximum continuous output of
2 kW at 25 mph and starts charging in a 10 mph wind. The
standard 13 ft propeller is designed to withstand a maximum
windspeed of 80 mph, a special short propeller, 10.5 ft in diameter will withstand winds of up to 120 mph but will give a
lower output than the larger propeller at given windspeeds.
This 80 mph windspeed limit is the only objection to the Dunlite, I do know of one machine which was destroyed by winds
in excess of 90 mph, but this is only a problem where such
high wind speeds are experienced.

The cost of the unit is 2000 Australian dollars, but don't
forget that importation will add considerably to the cost. Also
remember that if you purchase a Dunlite from an experienced
Agent you will also receive the benefit of his experience, and
that is worth a lot. Dunlite expect to manufacture a 5 to 6 kW
unit shortly and its expected cost is 3000 dollars Australian.

Dwyer Instruments Inc.,
Box 373, Michigan City, Indiana 46360, USA.

Manufacture the Dwyer Wind Meter and the Speed Indicator.
These are distributed in the UK by:

P.P. Controls Limited,
Crosslances Road, Hounslow, Middx.

Edmund Scientific Company,
222 Edscope Building, Barrington, New Jersey 08007, USA.

The Edmund Wind Wizard generates 600 watts at 25 mph.
This three bladed wooden propeller is geared to a 12 V Delco
generator. Regulated by a folding tail and "assembles in 10
minutes", it costs £475. It is not designed to withstand more

Photograph 37. The Edmund Wind Wizard.

than 50 mph and therefore is quite useless for any serious continuous use.

Elektro GmbH,
St. Gallerstrasse 27, Winterthur, Switzerland.

Elektro, a small company, have been quietly manufacturing wind generators for the past 37 years. All went well until about 1969 when suddenly everybody wanted to buy Elektros. The deluge of mail and orders arriving at their small workshop, where the machines are hand-built, meant that new people had to be employed who did not understand the fine art of 'wind craft' and so quality control slipped. About the same time Elektro allowed themselves to be pushed into premature manufacture of a 10 kW generator by an English company. The result was that the blades on the first batch of 10 kW mills broke. Elektro have suffered badly as a result. It must be said that Mr Schaufelberger has subsequently personally visited and de-bugged most of the faulty Elektros in Europe. I am satisfied that Elektro now have matters back in hand again and are intent upon improving their quality control.

After all that it should also be said that Elektro have many happy customers. Once an Elektro is working it tends to last a long time, 30 to 40 years. The following is a selection from their wide range of wind generators.

Maximum Output	Number of Blades	Cost Swiss Francs
600 watts	2	4300
1200 watts	2	5200
5/6000 watts	3	9500
8/10,000 watts	3	15000

Photograph 38. High in the Swiss Alps, a direct drive Elektro.

The prices include export packing, control panel, voltage regulator, but do not include tower, masthead, or various other accessories. Importation will also add to the cost. It is frequently best to purchase Elektros from a 'local' agent, provided that there is one, their experience will be helpful.

Elektro also sell suitable towers, batteries and inverters. They also manufacture wind generators for heating only. Their blades are made of wood with the variable pitch governed by centrifugal weights. All models up to the 4 kW are direct drive. Field magnet and permanent magnet alternators are used. Write to Elektro for details, or send 40 Swiss Francs for their recently written manual "Thirty-seven years experience with Elektro windmills".

Enag S.A.
Rue de Pont – L'Abbe, Quimper, Finistere, France.

Reported to manufacture twin bladed wind generators with outputs from 180 to 2000 watts, but they seem reluctant to let anyone know about them!

Grumman Energy Systems,
4175 Veterans Memorial Highway, Ronkonkoma, New York 11779, USA.

GES is a division of the huge Grumman Corporation. The Grumman Windstream 25 is a three bladed downwind mill. The variable pitch aluminium blades are electronically controlled. Centrifugal speed governors are located at the tip of each blade. The drive from the propeller is geared-up to the alternator. The power options are 20 kW at 28 mph or 15 kW at 26 mph. The Windstream and its concrete tower are designed to withstand 130 mph winds with a safety factor of 1.5, automatic shutdown occurs at 50 mph. The costs are as follows:

Windstream 25   £11,500
complete with control panel for battery charging or for operation with 15 or 20 kW Gemini inverter, and lightning cond-
uctor.

Tower   45 ft   £ 1125
Steel reinforced concrete tower.
    Grumman's first customer is the New York State Energy
Research and Development Authority who use the Wind-
stream to power a farm and they monitor its performance.

Photograph 39. Grumman Windstream 25.

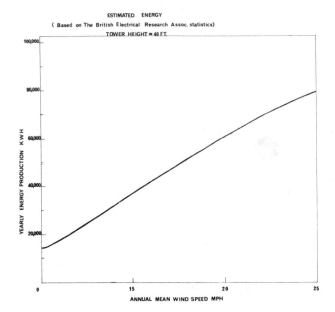

ESTIMATED ENERGY
(Based on The British Electrical Research Assoc. statistics)
TOWER HEIGHT = 40 F.T.

Figure 18. Grumman Windstream 25 estimated output.

Helical Power Limited,
36 Fontwell Drive, Glen Parva, Leicester LE2 9ML, UK.

The Helical rotor, as described earlier under Windmills, consists of two vertical blades forming a helical twist. Not yet
commercially available, though production of a 1 and 2 kW
model is envisaged.

Helion Inc.,
Box 445, Brownsville, California 95919, USA.

Helion is a non-profit research/education organisation started
many years ago by Jack Park, who is also the technical editor
of Wind Power Digest (see bibliography). Helion publish the
12/16 Construction Plans for the home construction of a three

112

bladed downwind propeller turbine. The 12 footer has a maximum output of 2 kW at 25 mph. A kit based on these plans is marketed by Topanga Power and a finished model is manufactured by Kedco Inc., both listed under Resources.

Helion manufacture a DC watt meter which is useful for monitoring a DC wind electric system. The meter costs £56 and they have also designed an Energy Source Analyser. This instrument measures and records wind speeds on 10 recorder channels, measures total solar insolation, and provides total time indication. The price for this miniature meteorological station is £685.

The home of Helion Inc. is a semi-autonomous ranch and they are now involved in the design of similar projects for others.

In general Helion supply a broad based renewable energy service, with a background of wind power.

Inverter Manufacturers

This is not a complete listing and therefore enquire locally for further manufacturers. A complete listing is given in the Electrical Research Association's yearly handbook which is available through libraries.

BKB Electric Motors Limited,
St George's Works, Camden Street, Birmingham.

Manufacture inexpensive rotary inverters up to 3 kW output.

Brunlec Appliances Limited,
Cowley Mill Road, Uxbridge, Middx.

200 and 500 watt square wave inverters.

Chelsfield Products Limited,
19 Broad Walk, Orpington, Kent.

300 watt square wave inverter.

Valradio Limited,
Browells Lane, Feltham, Middx.

Manufacture a wide range of inverters with outputs ranging up to 750 watts.

Kedco Inc.,
9016 Aviation Boulevard, Inglewood, California 90301, USA.

Kedco manufacture a family of four wind generators based on Park's three bladed downwind plans. Models 1200 (12 ft diameter) and 1600 (16 ft diameter) are both rated at 1200 watts maximum for battery charging applications, while models 1210 and 1610 are supplied with 2000 watt DC generators of permanent magnet design for synchronous inverter operation and for direct heating applications.

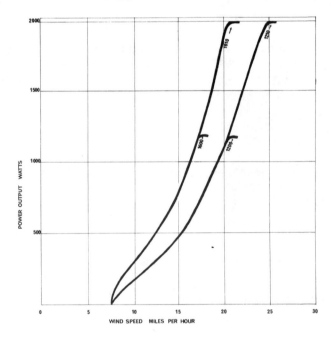

Figure 19. Output chart for the Kedco wind generator range.

114

Photograph 40. Kedco model 1600, 16ft diameter, 1200 watts at 18 mph.

The fundamental difference between the 1200 and 1600 series is the blade diameter. The increase in diameter from 12 to 16 feet nearly doubles the energy yield. Kedco offer, for a nominal price, a series of notes explaining the most appropriate ways to use their wind generator range.

All models have aluminium blades and the following features: automatic blade feathering, ground shut-off and reset cables, automatic vibration sensing shut-off and one year's warranty.

Prices

Model 1200	£1300
Model 1210	£1480
Model 1600	£1650
Model 1610	£1825

For export packing add a further £65

Low Energy Systems,
3 Larkfield Gardens, Dublin 6, Ireland.

Low Energy Systems have developed a vertical axis sailwing mill, suitable mainly for water pumping and other mechanical purposes. Its simple construction and low-speed output is comparable with the horizontal-axis Cretan mill.

The rotor consists of two or more sailwings. Each sailwing is formed from a rigid spar which is positioned at the leading edge of the sail. To this spar two or more rigid ribs are attached at right angles. The trailing edge of the sailwing is held in tension between the ends of the spars. The surface of the sailwing is made from cloth.

116

Photograph 41. Low Energy Systems vertical axis wind rotor.

In operation the sailwing takes up an aerofoil shape with a concave surface facing into the wind. During one complete revolution of the rotor the sailwing switches the concave surface from one side to the other automatically.

Unlike the Darrieus rotor it is self-starting. Its estimated efficiency of 30 per cent will shortly be tested in a wind tunnel. Home construction plans for this mill, suitable for water pumping and other mechanical purposes, are available from Low Energy Systems for £1.25 postpaid.

Lubing-Maschinenfabrik,
Ludwig Bening, 2847 Barnstort, Postfach 110, West Germany.

Lubing produce about fifty different sizes of wind driven water pumps and only one kind of wind generator, a downwind horizontal axis machine. The Lubing mill, rated at 400 watts in a 27 mph wind, has been steadily improved over the past 25 years of manufacture. Now, it is safe to say that the mill has achieved a reputation for robust design and reliability.

The six bladed Lubing propeller is shown in photograph 42. Three small, fixed pitch blades start the mill at 9 mph,

Photograph 42. The six-bladed Lubing machine.

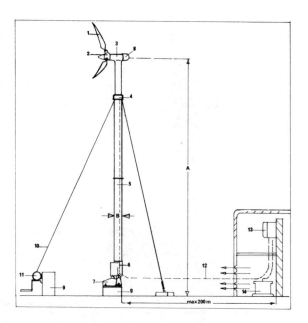

Figure 20. Lubing mounted on a folding tubular tower.

1. Rotor blade.
2. Rotor.
3. Crankcase section.
4. Rotating track ring.
5. Aluminium mast.
6. Swivel carriage housing.
7. Base of mast.

8. Generator.
9. Foundation.
10. Wire cable.
11. Manual cable winch.
12. Electric cable.
13. Switch box.
14. Batteries.

producing 24 watts. The three larger variable pitch blades produce the bulk of the power at higher windspeeds. All the blades are made of epoxy resins reinforced with glass fibre.

A centrifugal governor, fitted to each of the variable pitch blades, prevents them from exceeding the maximum speed of 600 rpm.

Transmission of power from shaft to the brushless alternator is by means of a two-stage oil-bath gear with a ratio of 5.5:1.

Output from the alternator is converted from AC to 24

volts DC at the control panel. The electronic control regulates the charging of the batteries automatically, when battery voltage of 28.5 volts is reached the charging current is cut off.

The price is high, 5537 German Marks (DM) for the basic unit with a stub tower (3 ft) and control panel.

The Lubing is also sold complete with an aluminium tubular tower in three sizes. The tubular tower is easy to bolt in place and has the added advantage of being hinged at the base which enables the owner to raise and lower the mill by means of the winch provided. This makes servicing and the annual gear-box oil change a relatively simple job.

Lubing wind generator with 23 ft tower	DM 7481
Lubing wind generator with 33 ft tower	DM 7895
Lubing wind generator with 42.5 ft tower	DM 8329

Joseph Lucas (Sales and Service) Limited,
Great Hampton Street, Birmingham.

Manufacture generators suitable for use with a windmill. Also manufacture batteries, 12 V fluorescent lights, water pumps and other 12 V equipment.

Metway Electrical Industries Limited,
Canning Street, Brighton, Sussex.

Manufacture 12 V immersion water heaters.

Natural Power Inc.,
Francestown Turnpike, New Boston, New Hampshire 03070, USA.

Natural Power, "specialists in electronics for the renewable energy industry", manufacture a remarkable selection of anemometers and wind monitors in general. If there is anything you want to know about the wind one of their instruments

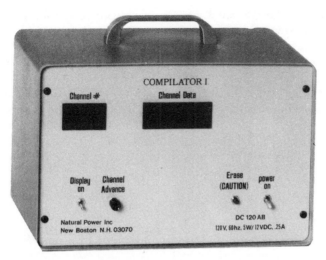

Photograph 43. Natural Power Wind Speed Compilator.

will do it. They also manufacture control panels, dynamic loading switches and alternators for use with wind power. The 1.5 kW alternator costs £165 for 12 and 24 V models. The Natural Power Tower Inc. Octahedron Module Tower is self-supporting and based on a design by Windworks. It is fully hot-dipped galvanised and can take a wind loading of up to 125 mph with a wind generator up to 20 ft rotor diameter. It can be erected from the ground up thus eliminating the need for a crane or gin pole.

25 ft tower	£415	51.5 ft tower	£830
34 ft tower	£530	70 ft tower	£1085
42.5 ft tower	£660	89 ft tower	£1485

Tower tops £65, with bearings £130.

Natural Power also have a bookshop and manufacture solar equipment.

Natural Power Systems Limited,
40 Sanderstead Road, South Croydon, Surrey.

With over two years of testing behind them Natural Power now feel ready to manufacture the Vortex Aero Generator. The Vortex is a twin bladed upwind mill rated at 3.5 kW in a 22 mph wind (250 watt at 7 mph). The fibre glass blades are 'spun' to form a web structure. Drive from the propeller is taken through a differential gearbox to the 3 phase alternator. A wind pressure switch automatically operates a differential hydraulic braking system when wind speeds exceed 35 mph approximately – brake release is automatic when the wind speed falls. The Vortex has been tested at wind speeds in ex-

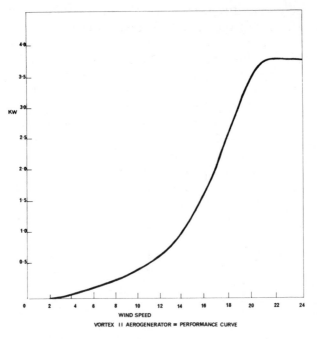

VORTEX II AEROGENERATOR ≈ PERFORMANCE CURVE

cess of 110 mph. Current from the mill can be used for battery charging or direct heating purposes. The Vortex Aero Generator costs £2500.

Natural Power also manufacture solar panels.

122

North Wind Power Company, Inc.,
Box 315, Warren, Vermont 05674, USA.

North Wind is best known for its rebuilt Jacobs Wind Electric plants. Jacobs manufactured and sold tens of thousands of wind generators in the period 1931 - 1956. In that time they have proved to be sturdy machines 'built to last a lifetime'. North Wind, who have a stock of old Jacobs, use them as the base for their completely rebuilt machines. Each generator is reconditioned, armature rebuilt, new brushes and bearings installed, and finally tested. Reconditioned or new propellers are supplied with either fly-ball or blade-control governors.

Jacobs are driven by three bladed, 14 ft diameter propellers are made from aircraft quality Sitka Spruce. The propeller direct drives a special slow speed 6 pole DC generator which gives its full output at a mere 350 rpm in windspeeds of under 20 mph. Each generator weighs over 450 lbs, the field coils alone contain close on 75 lbs of copper. The carbon brushes used with the Jacobs DC generator are known to last up to 20 years, a vast improvement on most brushes which rarely last more than one year. The direct drive from propeller to generator avoids the cost and trouble of gears. Indeed there are many who believe the Jacobs, especially the later 3 kW model, to be the best and most durable wind generator ever manufactured. Certainly one can say that these machines have never been infected with the modern industrial disease known as 'built-in obsolescence'.

Prices

Jacobs 2 kW, 32 volt	£1250
Jacobs 2 kW, 110 volt	£2000
Jacobs 3 kW, 32 volt	£1825
Jacobs 3 kW, 110 volt	£2625

The two 110 volt models are supplied with new propellers and blade-control governor.

North Wind also sell new and used self-supporting and

guyed towers. Their catalogue/information package also contains details of their extensive range of solid state and rotary inverters, and batteries. It is good value at £1.50. North Wind also sell light bulbs and motors which operate on 32 volt DC.

Apart from dealing in Wind Power, North Wind offer a complete integrated energy service.

Pinson Energy Corporation,
Box 7, Marstons Mills, Massachusetts 02648, USA.

Pinson Energy have just started to manufacture the Cyclo-turbine, a self-starting variable pitch vertical axis wind generator. Herman Drees and his team of highly qualified and enthusiastic helpers have been building and testing the Cyclo-turbine (see photograph under Windmills) for years, as all windmills should be before they are manufactured and sold. The blade length is 8 ft and the diameter is 12 ft. It will start in a 5 mph wind, and gives a maximum of 4 kW at 30 mph (2 kW at 24 mph). The unit, complete with a 30 ft tower, costs £2850.

P.I. Specialist Engineers Limited,
The Dean, Alresford, Hants, UK.

Photograph 44. Variable pitch vertical axis rotor.

124

Currently preparing to manufacture a variable pitch vertical axis wind generator, see figure 9 under Windmills. The unit will have two wooden blades rotating on an aluminium arm. It will generate 500 watts in a 15 mph wind. The expected price is about £2000. The turbine is based on the work of Dr Peter Musgrove at the University of Reading. An electric start will be used as the rotor is not self-starting.

Sencenbaugh Wind Electric,
2235 Old Middlefield Way, Mount View, California 94306.

Sencenbaugh Wind Electric was started in 1972 by Jim Sencenbaugh. Apart from acting as an agent for Dunlite, they have developed their own product range as follows:

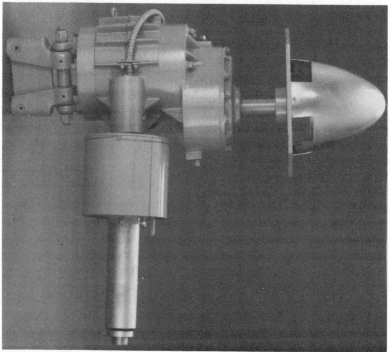

Photograph 45. Sencenbaugh 1000-14 close-up of hub, gear and alternator.

1. The Sencenbaugh Wind Generator Model 1000-14 is an up-wind horizontal axis machine and is rated at 1 kW in a 22mph wind. It has a three bladed propeller, 12 ft in diameter, constructed of machine carved Sitka Spruce with a bonded copper leading edge. The propeller speed is 175 rpm at cut-in and 290 rpm at maximum output.

Transmission is through a helical gear with a 3:1 ratio, over-designed to carry only 25 per cent of rated capacity at full generator output. Gear and alternator are both sealed in a cast aluminium body. Maximum continuous output at 14 V DC is 1000 watts at 22-23 mph, peak output is 1200 watts (180 watts at 10 mph). The 3 phase 6 pole alternator is slow speed.

The cut-in and charge rate is electronically controlled by a reliable solid state speed sensor and voltage regulator pioneered by Sencenbaugh in 1973.

Propeller overspeed control is provided by utilising the increase in wind pressure on the propeller and propeller thrust to furl the propeller downwind in high wind speeds. The foldable tail automatically begins to fold because of this force in winds above 25-30 mph. Reopening of the tail is automatic, being gravity fed. Neat! This system provides a simple yet positive method of overspeed control. It also protects windplant from gales and airborne debris. Maximum design windspeed is 80 mph and the turbine is built to a safety factor of 1.5. The price is £1500 and this includes the control panel but not a stub tower. It is available in 12 or 24 volt models and for a small additional cost a special marine model is available. This model, with stainless steel hardware, is highly recommended for coastal areas.

2. The Sencenbaugh Model 500-14 has the same characteristics except for the following:
The 500-14 has an output of 500 watts at 25 mph and a peak output of 600 watts (50 watts at 12 mph). The propeller diameter is only 6 ft.

Basically, this machine is designed for use in severe climates with high average windspeeds. As such it has a small propeller, direct drive (no gears) and is designed to withstand a

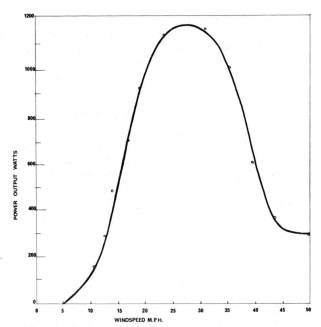

Figure 21. Output chart for Sencenbaugh Model 1000-14.

Photograph 46. Sencenbaugh model 500-14. 500 watts in a 24 mph windspeed.

127

maximum windspeed of 120 mph with a safety factor of 1.5. The propeller speed at cut-in is 280 rpm, and at maximum output 1000 rpm. The price of this model is £1250.

Solar Plexus,
West Street, Hadley, Massachusetts 01035, USA.

Solar Plexus make urethane polymer blades for the home builder. Blades, each 5 ft long, are strengthened with a metal spar running down the middle. Used in sets of three, they will produce 1.5 kW in a 20 mph wind with a tip speed ratio of 5:1. A set of three blades costs £52.

Topanga Power,
Box 712, Topanga, California 90290, USA.

Topanga manufacture a home construction kit based on the Helion 12/16 plans. The kit is sold in a semi-finished state or as individual components. Tasks left to the builder are final polishing and fitting of brushes, painting carriage and tower adaptor, bolting components into place, and pop-riveting the blades together. This work takes about two man days.

The resulting three bladed downwind machine can be supplied with a 12 or 16 ft diameter propeller. Due to alternator limitations, the maximum power output will be the same with either set of blades, but the advantage of longer blades is that full power will come at a much lower windspeed. The long blades deliver full power at 18 mph, and the short blades at 21 to 22 mph. Maximum power from the 12 volt alternator is 1200 watts and from the 24 volt model is 1800 watts.

The blades are made from aluminium and the transmission is via a gearbox with an 8.6:1 ratio. The weighted, self-feathering blades will continue to develop full power even when partially feathered. Ground operated shut off is available as an optional extra. Price £800 excluding stub tower.

See the entry under Kedco Inc., for similar fully manufactured and assembled model.

TWR Enterprises,
355 South Riverside, Rialto, California 92376, USA.

Sells a range of plans and home-built wind generator kits, send £1.00 for details.

Unarco-Rohn,
6718 West Plank Road, Box 2000, Peoria, Illinois 61601, USA.

For several years Rohn have supplied both self-supporting and guyed towers for use with Elektro, Dunlite, Jacobs, Sencenbaugh, Aero Power, and many other wind generators. Rohn towers also come complete with top section suitable for each type of wind generator.

Guyed tower suitable for Sencenbaugh mills (with top section)

40 ft tower	£285
60 ft tower	£342
80 ft tower	£495

Guyed tower suitable for Dunlite and Elektro (with top section)

40 ft tower	£342
60 ft tower	£528
80 ft tower	£720
100 ft tower	£915

Self-supporting tower with top section for Dunlite and Elektro.

40 ft tower	£660
60 ft tower	£915
80 ft tower	£1260
100 ft tower	£1675

Photograph 47. The Ballyhale Wind Pump.

Water Pumping Windmill Manufacturers

Southern Steel Works Limited,
Ballyhale, Co. Kilkenny, Ireland.

Manufacture the Ballyhale Wind Pump range of multi-bladed water pumping mills.

Sparco Pumps,
A/S Naesbjerg, 6800 Varde, Denmark.

Manufacture two small twin-bladed inexpensive water pump mills, one a diaphragm pump and the other a piston pump.

Photograph 48. The Sparco water pumping mill.

Wakes and Lamb Limited,
Millgate Works, Newark, Notts, UK.

Manufacture the Newark multi-blade water pump mill.

Wyatt Brothers Limited,
Whitchurch, Salop, UK.

Manufacture the Climax multi-blade mill.

Winco, Dyna Technology Inc.,
Box 3263, Sioux City, Iowa 51102, USA.

The Winco Wincharger 200 watt has been around for a long
time. The 6 ft propeller direct drives a fairly sturdy 4 pole
generator, both generator and slip ring brushes do need to be
replaced occasionally. It is available in 12, 24, 28, 32 and 36
volts, and comes complete with instrument panel and a 10 ft

Photograph 49. Winco Wincharger 200 watts.

stub tower for about £350. Maximum output is 200 watts at 23 mph.

Average usable kWh per month.

10 mph average	20
12 mph average	26
14 mph average	30

The Wind Energy Supply Company Limited,
Bolney Avenue, Peacehaven, Sussex, U.K.

WESCO is the result of the combined forces of two advanced technologies, one specialising in helicopter rotors and the other in control systems. Together they plan to manufacture a full range of wind machines adapted to various uses.

The gaunt machine shown in photograph 51 is the Oleo Hydraulic, used for direct heating. The 60 foot propeller is supported by a 45 foot steel tower. The two downwind blades are made of fibreglass with steel spars and the basic design is derived from many years of experience with helicopter rotors. The variable pitch rotor will produce about 7 kW at 10 mph windspeeds and a maximum of 190 kW at 30 mph. On a site with an average annual windspeed of 11 mph the output of heat energy is likely to be about 150,000 kWh a year. Weighted flaps on both blades prevent the propeller from exceeding a maximum of 120 rpm.

The propeller shaft and initial gear-up are made of a rear half-shaft from one of those huge earth-moving trucks. The mechanical energy from the shaft is then converted to heat energy by hydraulics. The hydraulic oil is then pumped from from the tower head, through insulated pipes, to a heat exchanger in the greenhouse. There the heat is transferred to water circulating through the greenhouse heating system.

Even though WESCO are happy with the results of their initial tests, I remain a little wary of twin-bladed downwind mills. My wariness is based on the apparent failure of the similar NASA 100 kW mill. The NASA mill suffered from excessive vibration as the blades passed through the 'wind shadow'

133

Photograph 50. Detail of the WESCO mill showing speed govenors on the blades.

cast by the tower. The next few months of tests should show whether my doubts are justified.

Apart from this mill WESCO plan to manufacture a com-

Photograph 51. WESCO's 60ft diameter Oleo Hydraulic windmill for direct heating.

plete range of wind machines with outputs ranging from 0.5 to 200 kW of electricity or heat equivalent. The propeller diameters in their basic range will be 10, 16, 23 and 60 foot. Apart from their novel hydraulic direct heat system, WESCO are also involved in developing wind systems in conjunction with chemical heat storage, solar collectors and heat pumps.

WESCO have at least two other mills on their drawing boards. One is a three-bladed upwind mill sold as a home-assembly kit and used for domestic wind electric systems. The second is a remarkable wind electric system driven by a single-bladed propeller. Yes, half a propeller and self-starting too! The idea is that with such a low blade solidity high speed is attained, resulting in a wind generator giving its maximum output of 500 watts in an 11 mph windspeed! To complement their range of wind generators WESCO have developed an interesting electronic control system (patents pending). In operation this control system will pass current, at a pre-selected voltage and frequency, direct to the electric load thus by-passing the batteries and therefore reducing the quantity of expensive batteries required. When the output from the wind generator exceeds the load the excess is fed to batteries which in turn supply power when output is less than load. The enormous benefit of this electronic control system became obvious when used with a wind generator giving its maximum output in an 11 mph windspeed.

Finally it must be said that I came away from a meeting with WESCO much impressed with their continued enthusiasm for wind power, despite the years of hard and unrewarding work. I was also glad to see how their enthusiasm is backed by considerable technical expertise. WESCO are worth watching, and good luck to them.

Wind Power Systems,
8871, A Balboa Avenue, San Diego, California 92123, USA.

Wind Power Systems have developed a unique wind generator operated by three propellers, the rims of which drive the alternator. The RD-4000 gives a maximum output of 4 kW in a 20 mph wind, with a cut-in speed of 7 mph. Output from the 3 phase brushless alternator is 110 volts DC. To protect the rotor in high winds all three propellers tilt upwards, and are hydraulically returned in normal windspeeds. The RD-4000 is not yet commercially available.

WPS have designed and built a Dutch four-arm mill which helps to power the Pea Soup Anderson motel-restaurant complex at Santa Nella. The 46 ft diameter propeller drives an alternator which generates about 8 kW in 20 mph winds. The windmill, first of its kind to be built for a long time, is capable of withstanding winds of up to 140 mph!

WPS manufacture a 'Windplant Energy Simulator' which, when using an anemometer, simulates the performance of a wind generator on any site where the simulator is sited. This gadget costs £400.

*Photograph 52. Wind Power
Systems RD - 4000.*

Windworks,
Box 329, Route 3, Mukwonago, Wisconsin 53149, USA.

Windworks is an engineering firm active in wind energy, power
conditioning and load management, advanced structural systems, and publishing. Perhaps their most interesting line is the
Gemini synchronous inverter. See 'Jacobs 2 kW Installation'
for details of operation.

For wind systems with capacities up to 20 kW synchronous inverters can be one sixth of the cost of conventional inverters per kilowatt capacity and may result in up to a 50 per
cent reduction in the capital cost of the entire system. With a
higher system efficiency and a lower capital cost for the same
generating capacity, the kilowatt hour cost is considerably
less than with battery/inverter systems. Because the nature of
the service provided by the utility, that of a storage medium
and current regulator, is different than their usual role, it will

137

be important to establish some initial agreement with the utility.

Windworks can supply Gemini inverters rated up to any capacity. The following gives an example of the costs:

4 kW maximum power conversion	£445
8 kW maximum power conversion	£828

Conversion efficiency at maximum power is 95 per cent. The singular disadvantage of the Gemini is its dependence upon the grid, but even so a battery/inverter can be installed if or when desired.

Windworks developed the octahedron tower now manufactured by Natural Power Inc. Their "25 ft Sail Windmill plans" are reviewed in the bibliography. They also publish a "Wind Energy Chart" which is a fine visual depiction of the chronology of wind power development, with general wind power data on the reverse, £2 from Windworks and their "Wind Energy Bibliography" is a good buy at £1.75.

WTG Energy Systems Inc.,
Box 87, One La Salle Street, Angola, New York 14006, USA.

WTG have just completed the erection of a 200 kW wind generator, the three bladed propeller has a diameter of 80 ft. The design is based on the Dutch Gedser mill, the most successful large wind generator ever built. The new mill will supply power to the residents of Cuttyhunk Island, Massachusetts. It will be interesting to see how this machine compares with the NASA 100 kW wind generator.

Zephyr Wind Dynamo Company,
Box 241, Brunswick, Maine 04011, USA.

The Zephyr Wind Dynamo is a three bladed downwind machine with a maximum output of 15 kW at 30 mph (500 watts at 10 mph). The low speed, direct drive alternator is specific-

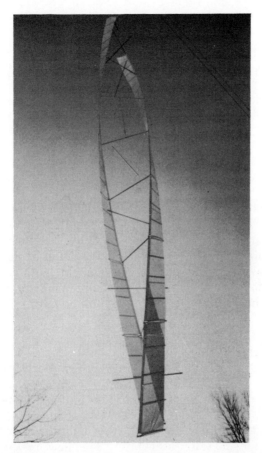

Photograph 53. Zephyr's experimental Tetra-helix wind machine.

ally designed for wind driven use and is of the permanent magnet type. The blades are a light weight composite of an injection moulded urethane foam body and coated Kevlar skin. Overspeed protection is by glide-out spoilers located near the blade tips. The maximum tip speed ratio is 6.5:1.

The price of the Zephyr, including control panel and 14 ft tower, is approximately £6850. This figure varies according to the intended use of the machine. Whilst Zephyr have sold three of their machines, two for test purposes, they still regard their new machine as an experimental one. Should the tests, and time, prove the design to be a good one, then confidence

Photograph 54. Zephyr Wind Dynamo with owner-built home in background.

will increase, as will production, thus causing a reduction in cost.

Zephyr have built a prototype of a rather unusual look-inf slant axis, wind machine called the Tetra-helix. The drawing outlines the idea. The whole structure is rigid under tension and will collapse easily without tension. Zephyr are currently testing their 2.5 kW prototype and the results should be interesting.

Zephyr also manufacture a smaller model, called the Tetra-helix S, and this is mainly intended for trickle charging batteries on small boats. The sails, made of Dracon sailcloth, describe a helical pattern about the torque delivery system, resulting in a self-starting, omni-directional rotor. Maximum output from the permanent magnet alternator is 7 watts in a 25 mph wind. The price is £145.

Agents

Alternative Energy Company,
Woodford, Co. Galway, Ireland.

Agent for Lubing, Winco and other generators.

Alternative Energy Services Limited,
2 Circular Road, Douglas, Isle of Man.

Agents for various wind generators.

D.D.Billsdon Engineering Limited,
88-90 London Road, Leicester.

Agent for Dunlite wind generators.

Cliffe Developments Limited,
Cliffe House, Village de Putron, Guernsey, Channel Islands.

Agent for Dunlite wind generators.

The Natural Energy Centre,
161 Clarence Street, Kingston Upon Thames, Surrey.

NEC have taken over from Conservation Tools and Technology Limited. They are agents for Winco and Elektro.

Bibliography

The most important and valuable up-to-date writing on wind energy is contained in the "Wind Power Digest", which is listed later.

Alternative Sources of Energy. A very fine journal with many excellent articles on wind power, including 'Martin Answers' by Martin Jopp. ASE is the journal for the home builder, it contains lots of solid, safe and intelligent advice. They have published many plans (and by far the best) for home builders. ASE No 24 (Feb '77) is an issue which features wind power particularly. Six issues yearly and well worth 16 dollars (US) from ASE, Route 2, Milaca, Minnesota, USA.

Brace Research Institute Publications.

How to construct a cheap wind machine for pumping water. A. Bodek. £1.00.

Performance test of a Savonius Rotor. Simonds and Bodek. £1.50.

Notes of the development of the Brace Airscrew Windmill as a prime mover. R. Chilcott. £0.50.

A simple electric transmission system for a free running windmill. Barton and Repole. £1.50.

Brace Research Institute
Macdonald College, Ste Anne de Bellevue, Quebec, Canada.

Catch the Wind. Landt and Lisl Dennis. A well written introduction to wind power. £4.50 from Four Winds Press,
50 West 44 th Street, New York, NY 10036, USA.

Do-it-Yourself Sail Windmill (Cretan) Plans. How to build a
12 ft diameter Sail Wind Generator, 200 watt output at
15 mph, maximum output 300 watts. Shown in the
photograph on page 30. Cost £0.60 from Centre for
Alternative Technology, Machynlleth, Powys, Wales.

Electric Power from the wind. Henry Clews. This booklet
describes the systems used with a 6 kW Elektro and a
2 kW Dunlite at the author's home, also makes interesting comparisons between the Dunlite and Elektro.
£1.50 from Enertech, Box 420, Norwich, Vermont 05055,
USA.

Energy from the Wind. Burke and Meroney. The bibliography
of all bibliographies on wind energy! Complete and ann-
otated. Costs £4.25 and there is also available a
First Supplement from 1975 to 1977 which costs £5.75. Both
are available from Publications, Engineering Research
Centre, Foothills Campus, Colorado State University.
Fort Collins, Colorado 80523, USA.

Food from Windmills. Peter Frankael. Describes the windmill
building activities of the American Presbyterian Mission
in Ethiopia. Contains a lot of good details on how to
build sail mills — mainly 11 ft diameter and used for
water pumping. Available from Intermediate Technology
Publications, 9 King Street, London WC2, England.

The Generation of Electricity by Wind Power. E.W.Golding.
Originally published in 1955, this remains a technical
masterpiece on wind power. Spon Limited, London.

Helion Model 12/16 Windmill Plans. See Helion entry under
Manufacturers for details.

The Homebuilt, Wind-generated Electricity Handbook. Michael Hackleman. The fact that this book is mis-titled does not detract from its brilliance as a complete guide to the rebuilding and installation of old Jacobs and Windchargers and any who have or intend to buy such machines (or any old wind generator) are well advised to buy this book at £4.50 from Earthmind, 5246 Boyer Road, Maripose, California 95338, USA.

Homemade Windmills of Nebraska. Erwin Barbour. This reprint originally published in 1898, describes how to build weird and wonderful windmills for pumping and sawing. Available from Farallones Institute, 15290 Coleman Valley Road, Oecidental, California 95465, USA.

The Journal of the New Alchemists. Journal No 2 contains details of how to build a sailwing rotor. Available at £3.50 from N.A.I., Box 432, Woods Hole, Massachusetts 02543, USA.

The Mother Earth News. Box 70, Hendersonville, North Carolina 28739, USA. Mother Earth News is the magazine for self-sufficiency, occasional articles on wind power, but we could do with more. Well worth £5.75 for 6 issues a year. They have also reprinted
Homemade Six-Volt Wind-Electric Plants which was originally published in 1939. Send MEN £1.00 for this.

The Princeton Sailwing Program. Dr T. Sweeney. A short report on the two bladed Sailwing developed at Princeton University. Send £1.50 to Forrestal Campus Library of Princeton University, Box 710, Princeton, New Jersey 08540, USA.

Pumping Windmill (Savonius). DIY Plan 3 from National Centre for Alternative Technology. Plans for adapting a 45 gallon drum into a water pumping rotor. £0.60p

Rain, 2270 N.W. Irving, Portland, Oregon 97210, USA. A fine monthly 'Journal of Appropriate Technology'. April '77 issue contained an excellent article on the 200 kW Dutch Gedser mill; a mill much praised for its low cost and suitability for local manufacture. Rain subscription costs £5.75 and a single issue costs £0.75.

> "As you can see the automatic shut-off device for the Gedser mill is a pipe that comes up from the floor, and on top of the pipe is a cup in which sits a heavy over-sized ball. That ball is connected by a string to an old fashioned Square D type switch on the wall. When the tower starts to vibrate the ball rolls out of the cup and the string pulls the switch to stop the machine."

Reinforced Brickwork Windmill Tower. A.B. Bird. Mainly on designing and building brickwork towers, also contains a section on how to build a 40 ft diameter Sail Windmill, photographed on page 32. It costs £1.15 from Structural Clay Products Limited, 230 High Street, Potters Bar, Herts, UK.

Simplified Wind Power Systems for Experiments. Jack Park. The core of this book contains data, of interest to home builders, on the aerodynamic, structural and mechanical design of wind driven propellers. £4.50 from Jack Park, Box 4301, Sylmar, California 91342, USA.

Wind Energy Bibliography. Windworks. A compilation of books, articles and papers. Sections include: Wind, windmills, aerodynamics, electrics, towers, storage, conversion, hydrogen and commercial units. As much of a bibliography as one would need, and well worth £1.80.

Wind Engineering, edited by E. Mowforth, University of Surrey. Multi-Science Publishing Company. A technical/academic quarterly journal on wind energy. Subscription £20 per annum for 4 issues with reduced rates to individuals.

Wind Power Digest. WPD is essential reading for the potential wind generator owner. Not only is it essential but it is also a most enjoyable read. Mike Evans is the editor and Jack Park is the technical editor. Joe Carter does some brilliant reporting on, and interviews with, prominent wind power people. All in all the Digest successfully reflects the excitement and enthusiasm of the wind power movement. At a yearly (4 issues) subscription rate of £3.50 you cannot go wrong. Back issues cost £1.25 and it is published by Jester Press, 54468 CR31, Bristol, Indiana 46507, USA.

Windustries. A fine regional quarterly newsletter, mainly on wind power. Subscription is £5.75 yearly (£8.75 for institutions). It is available from Great Plains Windustries, Box 126, Lawrence, Kansas 66044, USA.

Wind and Windspinners. Michael Hackleman. The first half of this book looks at the electrics of wind systems, the second half is on how to build Savonius rotors. It is definitely the best book on the Savonius rotor. From: Earthmind, 5246 Boyer Road, Mariposa, California 95338, USA.

Windy Ten Dutch Windmill Plans. For those who want to build a Dutch Four-arm, propeller diameter 8 ft. It can generate up to 500 watts if coupled to a generator. Suitable for the serious romantic only. Plans cost £9.50 from Edmund Scientific, 1006 Edscorp Building, Barrington, New Jersey 08007, USA.

25 foot Sail Windmill. Detailed design manual for a 6 sail mill and a 42 ft octahedron module tower with a platform for sail reefing, intended for mechanical purposes or for gearing up to generate electricity. Typical power outputs are as follows:

5 mph windspeed	0.1 hp
10 mph windspeed	0.8 hp
15 mph windspeed	2.8 hp
20 mph windspeed	6.7 hp

It requires manual reefing in winds over 20 mph. Prepared for Brace Research Institute by Windworks. The plans are available for £15 from Windworks, Box 329 Route 3, Mukwonago, Wisconsin 53149, USA.

The following four books all include articles on wind, water, solar, biofuels and integrated energy designs:

Energy for the Home. Peter Clegg. Garden Way Publishing, Vermont, USA.

Energy Primer. Portola Institute. A new edition of this book is due to be published mid '78 by Dell Publishing, USA and Prism Press, UK.

Other Homes and Garbage. Lockie, Masters, Whitehouse and Young. Sierra Club Books. 1975.

Producing your own Power. Edited by C. Stoner. Rodale Press. 1975.